Navigating the Quantum Realm

A Journey into the Heart of Physics

Table Of Contents

Chapter 1: Introduction to Quantum Physics
- Defining Quantum Physics
- Historical Background: From Classical to Quantum
- Key Players in the Development of Quantum Theory

Chapter 2: The Wave-Particle Duality
- Wave-Particle Duality: A Fundamental Concept
- The Double-Slit Experiment
- Understanding Quantum Superposition

Chapter 3: The Uncertainty Principle
- Heisenberg's Uncertainty Principle
- Implications of Uncertainty
- Measurement Problem and Quantum Decoherence

Chapter 4: Quantum Mechanics Basics
- Wave Function and Probability Amplitude
- Schrödinger Equation
- Operators and Observables

Chapter 5: Quantum States and Quantum Entanglement
- Quantum States: Pure vs. Mixed
- Entanglement: Spooky Action at a Distance
- Bell's Theorem and Tests of Quantum Entanglement

Chapter 6: Quantum Mechanics in Practice
- Quantum Tunneling
- Quantum Computing
- Quantum Cryptography

Chapter 7: Quantum Field Theory
- Introduction to Quantum Field Theory
- Quantization of Fields
- Particle Physics and Quantum Field Theory

Chapter 8: Quantum Electrodynamics
- Fundamentals of Quantum Electrodynamics
- Renormalization: Dealing with Divergences
- Feynman Diagrams and Quantum Electrodynamics

Chapter 9: Quantum Chromodynamics
- The Strong Force and Quantum Chromodynamics
- Quarks and Gluons

- Asymptotic Freedom and Confinement

Chapter 10: Quantum Gravity and Unified Theories
- Challenges of Quantum Gravity
- String Theory
- Loop Quantum Gravity and Other Approaches

Chapter 11: Quantum Cosmology
- Quantum Origins of the Universe
- Inflationary Cosmology
- Quantum Cosmological Models

Chapter 12: Quantum Philosophy and Interpretations
- Copenhagen Interpretation
- Many-Worlds Interpretation
- Pilot-Wave Theory and Other Interpretations

Chapter 13: Applications of Quantum Physics
- Quantum Optics
- Quantum Sensing and Metrology
- Quantum Biology

Chapter 14: Future Directions in Quantum Physics
- Quantum Technologies on the Horizon
- Challenges and Opportunities
- The Future of Quantum Research

Chapter 15: Conclusion: Embracing the Quantum World
- Recapitulation of Key Concepts
- Importance of Quantum Physics in Modern Science and Technology
- The Beauty and Mystery of the Quantum Realm

Appendix: Glossary of Quantum Terms

Chapter 1: Introduction to Quantum Physics

Quantum physics is a branch of physics that deals with the behavior of matter and energy at the smallest scales, where the very nature of reality seems to defy our common-sense understanding. In this introductory chapter, we will explore the fundamental concepts of quantum physics and the historical journey that led to its development.

Defining Quantum Physics

Quantum physics, also known as quantum mechanics, is a theoretical framework that describes the behavior of particles at the atomic and subatomic levels. Unlike classical physics, which deals with macroscopic objects, quantum physics deals with the microscopic world, where particles such as electrons, photons, and atoms exhibit behaviors that are radically different from what we observe in our everyday lives.

At the heart of quantum physics is the notion that particles, such as electrons and photons, can exist in multiple states simultaneously, a concept known as superposition. Additionally, the behavior of particles is inherently probabilistic, meaning that we can only predict the likelihood of certain outcomes rather than determine them with certainty.

Historical Background: From Classical to Quantum

The development of quantum physics was a revolutionary journey that challenged centuries-old notions of determinism and causality. It began in the early 20th century with the observation of phenomena that could not be explained by classical physics.

One of the key experiments that paved the way for quantum theory was the photoelectric effect, discovered by Albert Einstein in 1905. This phenomenon demonstrated that light behaves as both particles (photons) and waves, laying the groundwork for the wave-particle duality concept in quantum physics.

Another pivotal moment came with Max Planck's introduction of the concept of quantization in 1900 to explain the behavior of blackbody radiation. Planck proposed that energy is not emitted or absorbed continuously but in discrete packets, or quanta, which led to the birth of quantum theory.

Key Players in the Development of Quantum Theory

Several prominent physicists played crucial roles in the development of quantum theory, each contributing groundbreaking insights that revolutionized our understanding of the microscopic world.

- Niels Bohr: Known for his work on the structure of atoms and the development of the Bohr model, which introduced the idea of quantized electron orbits.
- Erwin Schrödinger: Developed the wave equation, known as the Schrödinger equation, which describes the behavior of quantum mechanical systems.
- Werner Heisenberg: Formulated the uncertainty principle, which states that the position and momentum of a particle cannot be simultaneously measured with arbitrary precision.

- Paul Dirac: Made significant contributions to quantum mechanics, including the formulation of quantum field theory and the prediction of antimatter.

As we embark on our exploration of quantum physics, we must be prepared to embrace the counterintuitive nature of the quantum realm, where particles can be in two places at once and where observation fundamentally alters the outcome of experiments. It is a world of uncertainty and probability, yet it holds the key to understanding the underlying fabric of reality at its most fundamental level.

Defining Quantum Physics

Quantum physics, often referred to as quantum mechanics, is a branch of physics that examines the behavior of matter and energy at the smallest scales, encompassing the realm of atoms, subatomic particles, and the fundamental forces that govern their interactions. At its core, quantum physics seeks to understand the underlying principles that govern the behavior of particles and systems in this microscopic domain.

The defining characteristic of quantum physics is its departure from classical physics, which primarily deals with the behavior of objects on macroscopic scales. Unlike classical physics, where objects follow deterministic laws and have well-defined properties, quantum physics introduces a level of uncertainty and probabilistic behavior at the quantum level.

One of the central tenets of quantum physics is the concept of quantization, which suggests that certain properties, such as energy, angular momentum, and electric charge, exist in discrete, quantized amounts rather than continuously. This notion challenges the classical idea of continuity and introduces the idea of discrete states and transitions between them.

Another fundamental aspect of quantum physics is the wave-particle duality, which asserts that particles, such as electrons and photons, exhibit both wave-like and particle-like behavior depending on the context of the experiment. This duality is exemplified by phenomena such as the double-slit experiment, where particles exhibit interference patterns characteristic of waves when passed through two slits, suggesting the existence of wave-like properties.

Quantum physics also introduces the concept of superposition, wherein particles can exist in multiple states simultaneously until they are observed or measured. This idea challenges our classical intuition, where objects typically have well-defined properties at all times.

Superposition lies at the heart of quantum computing and quantum information processing, where information can be encoded in quantum states to perform parallel computations.

Furthermore, quantum physics introduces the uncertainty principle, formulated by Werner Heisenberg, which states that there is a fundamental limit to the precision with which certain pairs of complementary properties, such as position and momentum, can be simultaneously known. This principle underscores the inherent limitations of measurement and observation at the quantum level.

In summary, quantum physics provides a theoretical framework for understanding the behavior of matter and energy at the smallest scales, where classical intuition breaks down, and probabilistic phenomena dominate. Its principles have profound implications for fields ranging from particle physics and cosmology to technology and information theory, shaping our understanding of the fundamental nature of reality.

Historical Background: From Classical to Quantum

The transition from classical to quantum physics marks one of the most significant revolutions in the history of science. This historical journey is characterized by a series of groundbreaking discoveries and paradigm shifts that challenged long-held beliefs about the nature of reality and laid the foundation for the development of quantum theory.

Classical Physics: The Newtonian Framework

Classical physics, based on the principles established by Isaac Newton in the 17th century, provided a remarkably successful framework for understanding the behavior of objects in the macroscopic world. Newton's laws of motion and his law of universal gravitation allowed scientists to predict the motions of planets, the behavior of falling objects, and the dynamics of mechanical systems with remarkable accuracy.

For centuries, classical physics reigned supreme, offering a deterministic view of the universe where the future state of a system could be precisely determined given its initial conditions and the forces acting upon it. This deterministic worldview seemed to provide a complete and comprehensive description of physical phenomena, leaving little room for uncertainty or ambiguity.

The Birth of Quantum Physics: Planck and the Quantum Hypothesis

The seeds of the quantum revolution were sown at the turn of the 20th century, with the discovery of phenomena that defied classical explanation. One of the pivotal moments came in 1900 when the German physicist Max Planck introduced the concept of quantization to explain the behavior of blackbody radiation.

Planck proposed that energy is not emitted or absorbed continuously but in discrete packets, or quanta, whose size is proportional to the frequency of the radiation. This revolutionary idea, known as the quantum hypothesis, laid the groundwork for the development of quantum theory and challenged the classical notion of continuous energy exchange.

The Photoelectric Effect and Einstein's Contribution

Another crucial breakthrough came in 1905 when Albert Einstein provided a theoretical explanation for the photoelectric effect, a phenomenon in which light shining on a metal surface ejects electrons from it. Einstein's analysis proposed that light consists of discrete packets of energy, or photons, whose energy is proportional to their frequency.

Einstein's explanation of the photoelectric effect not only provided compelling evidence for the existence of quanta but also introduced the concept of wave-particle duality, wherein particles such as photons exhibit both wave-like and particle-like behavior. This concept laid the foundation for the development of quantum mechanics and challenged the classical distinction between particles and waves.

The Birth of Quantum Mechanics: Bohr, Schrödinger, and Heisenberg

The true birth of quantum mechanics as a comprehensive theoretical framework came in the 1920s, with the contributions of pioneering physicists such as Niels Bohr, Erwin Schrödinger, and Werner Heisenberg. Bohr's model of the atom, based on quantized electron orbits, provided a conceptual framework for understanding atomic spectra and laid the groundwork for future developments in quantum theory.

Schrödinger's wave equation, formulated in 1926, provided a mathematical description of the behavior of quantum mechanical systems, allowing physicists to calculate the probability distributions of particles in various states. Heisenberg's uncertainty principle, formulated around the same time, introduced the idea that there are inherent limits to the precision with which certain pairs of complementary properties, such as position and momentum, can be simultaneously known.

Together, these developments marked a radical departure from classical physics and ushered in a new era of scientific inquiry characterized by uncertainty, probability, and the fundamental indeterminacy of the quantum world. The stage was set for further exploration and discovery, leading to the development of quantum mechanics as we know it today.

Key Players in the Development of Quantum Theory

The development of quantum theory involved the contributions of numerous pioneering physicists who challenged classical physics and revolutionized our understanding of the microscopic world. Here, we highlight some of the key figures whose insights and discoveries shaped the foundation of quantum mechanics.

1. Max Planck (1858–1947)

Max Planck is often regarded as the father of quantum theory for his groundbreaking work on blackbody radiation, which led to the introduction of the concept of quantization. In 1900, Planck proposed that energy is emitted and absorbed in discrete packets, or quanta, rather than continuously. This idea laid the groundwork for the development of quantum mechanics and earned Planck the Nobel Prize in Physics in 1918.

2. Albert Einstein (1879–1955)

Albert Einstein made several seminal contributions to the field of quantum mechanics, most notably his explanation of the photoelectric effect in 1905. Einstein proposed that light consists of discrete packets of energy, or photons, which helped establish the concept of wave-particle duality. Despite his foundational contributions to quantum theory, Einstein famously expressed skepticism about certain aspects of quantum mechanics, famously stating, "God does not play dice with the universe."

3. Niels Bohr (1885–1962)

Niels Bohr played a central role in the development of quantum mechanics, particularly with his work on the structure of the atom. In 1913, Bohr introduced the Bohr model of the atom, which incorporated the quantization of electron orbits and provided a framework for understanding atomic spectra. Bohr's model laid the foundation for future developments in quantum theory and earned him the Nobel Prize in Physics in 1922.

4. Werner Heisenberg (1901–1976)

Werner Heisenberg is best known for formulating the uncertainty principle, a fundamental concept in quantum mechanics. In 1927, Heisenberg proposed that there is a fundamental limit to the precision with which certain pairs of complementary properties, such as position and momentum, can be simultaneously known. This principle fundamentally altered our understanding of the behavior of particles at the quantum level and remains a cornerstone of quantum mechanics.

5. Erwin Schrödinger (1887–1961)

Erwin Schrödinger made significant contributions to quantum mechanics, particularly with his formulation of the wave equation in 1926. Schrödinger's equation describes the behavior of quantum mechanical systems and provides a mathematical framework for calculating the probability distributions of particles in various states. Schrödinger's wave mechanics, along with Heisenberg's matrix mechanics, formed the basis of modern quantum mechanics.

6. Paul Dirac (1902–1984)

Paul Dirac was a pioneering physicist who made profound contributions to the development of quantum mechanics, particularly with his formulation of relativistic quantum mechanics. In 1928, Dirac proposed an equation that unified quantum mechanics with Einstein's theory of special relativity, leading to the prediction of the existence of antimatter. Dirac's equation provided a theoretical framework for understanding the behavior of particles at high energies and earned him the Nobel Prize in Physics in 1933.

These individuals, along with many others, played pivotal roles in the development of quantum theory, challenging classical physics and reshaping our understanding of the fundamental nature of reality. Their insights and discoveries continue to inspire generations of physicists and have profound implications for our understanding of the universe.

Chapter 2: The Wave-Particle Duality

The wave-particle duality is one of the most intriguing and fundamental concepts in quantum physics, challenging our classical intuition about the nature of matter and radiation. In this chapter, we explore the fascinating phenomenon of wave-particle duality, its experimental evidence, and its profound implications for our understanding of the quantum world.

Exploring the Dual Nature of Matter and Radiation

In classical physics, particles and waves were traditionally considered distinct entities with separate properties and behaviors. However, the advent of quantum mechanics shattered this dichotomy, revealing that particles, such as electrons and photons, exhibit both wave-like and particle-like characteristics depending on the context of the experiment.

The Double-Slit Experiment: A Cornerstone of Quantum Physics

One of the most famous experiments demonstrating wave-particle duality is the double-slit experiment. In this experiment, a beam of particles, such as electrons or photons, is directed towards a barrier with two narrow slits. Surprisingly, when the particles pass through the slits and are detected on a screen behind the barrier, they create an interference pattern characteristic of waves, even when emitted one at a time.

Wave-Like Behavior: Interference and Diffraction

The interference pattern observed in the double-slit experiment is a hallmark of wave behavior, indicating that particles can interfere with themselves as they pass through the slits. This interference phenomenon occurs when waves overlap and either reinforce or cancel each other out, resulting in a pattern of alternating bright and dark fringes on the detection screen. Furthermore, particles exhibit diffraction, another wave-like phenomenon, as they bend around obstacles and spread out after passing through narrow openings. The degree of diffraction depends on the wavelength of the particles, highlighting the wave-like nature of their behavior.

Particle-Like Behavior: Photons and the Photoelectric Effect

While the wave-like behavior of particles is evident in experiments such as the double-slit experiment, particles also exhibit particle-like behavior in certain situations. For example, the photoelectric effect, discovered by Albert Einstein, demonstrates that light behaves as discrete packets of energy, or photons, when interacting with matter.
In the photoelectric effect, photons incident on a metal surface eject electrons from it, with the kinetic energy of the ejected electrons depending on the frequency of the incident light rather than its intensity. This phenomenon provided compelling evidence for the particle-like nature of light and contributed to the development of quantum theory.

Reconciling Wave-Particle Duality: The Copenhagen Interpretation

The wave-particle duality presents a profound challenge to our classical understanding of the nature of reality. According to the Copenhagen interpretation of quantum mechanics, proposed by Niels Bohr and Werner Heisenberg, particles exist in a state of superposition, where they can simultaneously exhibit wave-like and particle-like behavior until measured or observed.
When a measurement is made to determine the properties of a particle, its wave function collapses, and it manifests as a particle with well-defined properties. This interpretation underscores the inherently probabilistic nature of quantum mechanics and emphasizes the role of observation in determining the outcome of experiments.

Conclusion: Embracing the Paradox of Wave-Particle Duality

The wave-particle duality challenges our classical intuition and forces us to confront the paradoxical nature of the quantum world. Particles behave as waves and waves behave as particles, blurring the boundaries between these fundamental concepts and reshaping our understanding of the nature of matter and radiation.
As we delve deeper into the mysteries of quantum physics, we must embrace the wave-particle duality as a fundamental aspect of the quantum realm. It is through experiments and observations that we uncover the intricate interplay between waves and particles, revealing the richness and complexity of the quantum world.

Wave-Particle Duality: A Fundamental Concept

Wave-particle duality is a cornerstone of quantum mechanics, presenting a profound paradox that challenges our classical understanding of the nature of matter and radiation. In this section, we delve into the essence of wave-particle duality, exploring its historical origins, experimental evidence, and implications for our understanding of the quantum world.

Historical Origins

The concept of wave-particle duality emerged in the early 20th century as physicists grappled with the behavior of particles at the atomic and subatomic levels. The first hints of this duality came from experiments such as the double-slit experiment and the photoelectric effect, which demonstrated that particles, such as electrons and photons, exhibit both wave-like and particle-like behavior under different experimental conditions.

Experimental Evidence

The double-slit experiment stands as a paradigmatic demonstration of wave-particle duality. In this experiment, a beam of particles, such as electrons or photons, is directed towards a barrier with two narrow slits. Surprisingly, when the particles pass through the slits and are detected on a screen behind the barrier, they create an interference pattern characteristic of waves, even when emitted one at a time. This interference pattern indicates that particles can interfere with themselves, suggesting wave-like behavior.

On the other hand, the photoelectric effect provides compelling evidence for the particle-like nature of light. In this phenomenon, photons incident on a metal surface eject electrons from it, with the kinetic energy of the ejected electrons depending on the frequency of the incident light rather than its intensity. This behavior suggests that light consists of discrete packets of energy, or photons, exhibiting particle-like behavior.

Implications for Quantum Mechanics

Wave-particle duality challenges our classical intuition about the nature of reality, blurring the distinction between particles and waves and suggesting that the behavior of particles is inherently probabilistic. According to quantum mechanics, particles exist in a state of superposition, where they can simultaneously exhibit wave-like and particle-like behavior until measured or observed.

This superposition of states lies at the heart of quantum mechanics, giving rise to phenomena such as quantum entanglement and quantum tunneling. It also underscores the inherently probabilistic nature of quantum mechanics, where measurements yield probabilistic outcomes rather than deterministic ones.

Conclusion

Wave-particle duality is a fundamental concept in quantum mechanics, revealing the dual nature of matter and radiation and challenging our classical understanding of the universe. As we continue to explore the mysteries of the quantum world, wave-particle duality serves as a guiding principle, reminding us of the intricate interplay between waves and particles and the profound implications for our understanding of the nature of reality.

The Double-Slit Experiment

The double-slit experiment stands as one of the most iconic and thought-provoking experiments in the history of physics, providing compelling evidence for the wave-particle duality of matter and radiation. In this section, we delve into the intricacies of the double-slit experiment, its historical significance, and its profound implications for our understanding of the quantum world.

Setting the Stage: The Experimental Setup

In the double-slit experiment, a beam of particles, such as electrons or photons, is directed towards a barrier with two narrow slits, hence the name "double-slit." Behind the barrier lies a detection screen that records the arrival of particles.

Wave-Like Interference: A Surprising Pattern Emerges

When particles are emitted one at a time and allowed to pass through the slits onto the detection screen, something remarkable occurs. Instead of observing two distinct bands of particles corresponding to each slit, as one might expect in a classical scenario, an interference pattern emerges on the detection screen. This pattern consists of alternating bright and dark fringes, characteristic of wave interference phenomena.

Understanding Interference: Waves at Play

The emergence of an interference pattern suggests that the particles are behaving like waves as they pass through the slits. When waves from each slit overlap and interfere with each other, they either reinforce or cancel each other out, resulting in regions of constructive and destructive interference on the detection screen. The resulting pattern reflects the wave nature of the particles and provides compelling evidence for their ability to interfere with themselves.

Particle-Like Detection: The Observer Effect

Remarkably, even when particles are emitted one at a time, the interference pattern still emerges over time as more particles accumulate on the detection screen. However, the act of observing or measuring which slit each particle passes through disrupts the interference pattern, causing the particles to behave more like classical particles and exhibit a simple two-band pattern on the screen.

This phenomenon, known as the observer effect, highlights the delicate interplay between wave-like and particle-like behavior and underscores the role of measurement or observation in determining the outcome of the experiment.

Implications for Quantum Mechanics

The double-slit experiment epitomizes the wave-particle duality of matter and radiation, challenging our classical intuition about the nature of reality. It suggests that particles, such as electrons and photons, possess both wave-like and particle-like characteristics, depending on the context of the experiment.

Furthermore, the double-slit experiment underscores the inherently probabilistic nature of quantum mechanics, where measurements yield probabilistic outcomes rather than deterministic ones. It highlights the profound implications of observation and measurement on the behavior of particles at the quantum level.

Conclusion: Provoking Questions and Inspiring Insights

The double-slit experiment continues to provoke questions and inspire insights into the nature of reality and the fundamental principles of quantum mechanics. Its remarkable findings challenge our classical understanding of the universe and underscore the rich and complex nature of the quantum world. As we continue to explore the mysteries of quantum physics, the double-slit experiment remains a testament to the profound mysteries and discoveries that lie at the heart of quantum mechanics.

Understanding Quantum Superposition

Quantum superposition is a fundamental concept in quantum mechanics that lies at the heart of the wave-particle duality and the probabilistic nature of quantum systems. In this section, we explore the concept of quantum superposition, its implications, and its significance in understanding the behavior of particles at the microscopic level.

What is Quantum Superposition?

Quantum superposition refers to the ability of quantum systems to exist in multiple states simultaneously until measured or observed. This means that particles, such as electrons or photons, can occupy multiple positions, velocities, or other properties at the same time, rather than having a single well-defined state.

Schrödinger's Cat: A Thought Experiment

One of the most famous illustrations of quantum superposition is Schrödinger's cat, a thought experiment proposed by physicist Erwin Schrödinger in 1935. In this scenario, a cat is placed in a sealed box with a device that has a 50% chance of releasing a poison, based on the random decay of a radioactive atom. According to quantum mechanics, until the box is opened and the cat is observed, it exists in a superposition of being both alive and dead simultaneously.

The Mathematics of Superposition: Wave Functions

In quantum mechanics, the state of a particle is described by a mathematical construct known as a wave function. The wave function encapsulates all possible states that a particle can occupy, each with an associated probability amplitude. When a particle is in a state of superposition, its wave function is a combination of these possible states, weighted by their probability amplitudes.

Interference and Superposition

Quantum superposition leads to interference phenomena, where the wave functions of particles overlap and interfere with each other. This interference can result in constructive or destructive interference, amplifying or canceling out certain states, respectively. Interference plays a crucial role in many quantum phenomena, including the double-slit experiment, where particles create interference patterns on a detection screen.

Measurement and Collapse of the Wave Function

When a measurement is made to determine the state of a particle, the superposition collapses, and the particle manifests in a single, well-defined state. This process, known as wave function collapse, occurs due to the interaction between the quantum system and the measuring apparatus, leading to the emergence of a definite outcome.

Implications for Quantum Computing and Information

Quantum superposition is central to the potential of quantum computing and quantum information processing. By encoding information in the superposition of quantum states, quantum computers can perform parallel computations and solve certain problems more efficiently than classical computers. Furthermore, quantum cryptography relies on the principles of superposition to ensure secure communication.

Conclusion: Embracing the Paradox of Superposition

Quantum superposition is a fundamental aspect of quantum mechanics that challenges our classical intuition about the nature of reality. It underscores the probabilistic nature of quantum systems and the delicate interplay between wave-like and particle-like behavior. As we continue to explore the mysteries of the quantum world, quantum superposition remains a central concept that reshapes our understanding of the fundamental principles of physics.

Chapter 3: The Uncertainty Principle

The uncertainty principle, formulated by Werner Heisenberg in 1927, is a fundamental concept in quantum mechanics that imposes limits on our ability to simultaneously measure certain pairs of physical properties, such as position and momentum, with arbitrary precision. In this chapter, we delve into the intricacies of the uncertainty principle, its implications for quantum mechanics, and its profound implications for our understanding of the quantum world.

Introduction to the Uncertainty Principle

The uncertainty principle states that there is a fundamental limit to the precision with which certain pairs of complementary properties of a particle, such as position and momentum, can be simultaneously known. In other words, the more precisely we know one property, the less precisely we can know the other. This introduces an inherent uncertainty into the measurement process at the quantum level.

Heisenberg's Formulation

Werner Heisenberg originally formulated the uncertainty principle in the context of matrix mechanics, one of the early formulations of quantum mechanics. He derived the principle mathematically, showing that the product of the uncertainties in the measurement of position (Δx) and momentum (Δp) of a particle is bounded by a constant, known as Planck's constant (h-bar), divided by 2π: $\Delta x * \Delta p \geq$ h-bar / 2.

Implications for Quantum Mechanics

The uncertainty principle has profound implications for the behavior of particles at the quantum level. It implies that particles do not have well-defined properties, such as position and momentum, until they are measured or observed. Instead, particles exist in a state of superposition, where they can simultaneously occupy a range of possible states, each with an associated probability amplitude.

Heisenberg's Thought Experiment

Heisenberg illustrated the uncertainty principle with a famous thought experiment involving the measurement of the position and momentum of an electron using a microscope. He showed that the act of observing the electron with a microscope necessarily disturbs its momentum, leading to uncertainty in both position and momentum measurements. This thought experiment highlights the inherent limitations of measurement at the quantum level.

Applications and Interpretations

The uncertainty principle has far-reaching implications for various areas of physics, including quantum mechanics, quantum field theory, and quantum information theory. It plays a crucial role in understanding phenomena such as quantum tunneling, atomic and molecular structure, and the behavior of subatomic particles.

Conclusion: Embracing the Uncertainty

The uncertainty principle fundamentally alters our classical intuition about the nature of reality, introducing an inherent uncertainty into the fabric of the universe. It underscores the probabilistic nature of quantum mechanics and the delicate interplay between observation and measurement. As we continue to explore the mysteries of the quantum world, the uncertainty principle remains a cornerstone of quantum theory that reshapes our understanding of the fundamental principles of physics.

Heisenberg's Uncertainty Principle

Heisenberg's Uncertainty Principle, proposed by the German physicist Werner Heisenberg in 1927, is a fundamental concept in quantum mechanics that fundamentally changed our understanding of the behavior of particles at the atomic and subatomic levels. In this chapter, we explore the formulation, implications, and significance of Heisenberg's Uncertainty Principle.

Formulation of the Uncertainty Principle

Heisenberg's Uncertainty Principle states that there is a fundamental limit to the precision with which certain pairs of physical properties of a particle can be simultaneously known. In particular, it imposes limits on our ability to measure the position and momentum of a particle simultaneously with arbitrary precision.

Mathematically, the uncertainty principle is expressed as: $\Delta x * \Delta p \geq \hbar / 2$, where Δx represents the uncertainty in the position measurement, Δp represents the uncertainty in the momentum measurement, and h-bar is the reduced Planck's constant.

Implications for Quantum Mechanics

The uncertainty principle has profound implications for our understanding of quantum mechanics. It implies that particles do not have well-defined properties, such as position and momentum, until they are measured or observed. Instead, particles exist in a state of superposition, where they can simultaneously occupy a range of possible states, each with an associated probability amplitude.

Furthermore, the uncertainty principle highlights the inherently probabilistic nature of quantum mechanics, where measurements yield probabilistic outcomes rather than deterministic ones. It underscores the delicate interplay between wave-like and particle-like behavior and the limitations of measurement at the quantum level.

Heisenberg's Thought Experiment

Heisenberg illustrated the uncertainty principle with a famous thought experiment involving the measurement of the position and momentum of an electron using a microscope. He showed that the act of observing the electron with a microscope necessarily disturbs its momentum, leading to uncertainty in both position and momentum measurements. This thought experiment highlights the inherent limitations of measurement at the quantum level.

Applications and Interpretations

The uncertainty principle has far-reaching implications for various areas of physics, including quantum mechanics, quantum field theory, and quantum information theory. It plays a crucial role in understanding phenomena such as quantum tunneling, atomic and molecular structure, and the behavior of subatomic particles.

Conclusion: Embracing the Uncertainty

Heisenberg's Uncertainty Principle fundamentally alters our classical intuition about the nature of reality, introducing an inherent uncertainty into the fabric of the universe. It underscores the probabilistic nature of quantum mechanics and the delicate interplay between observation and measurement. As we continue to explore the mysteries of the quantum world, the uncertainty principle remains a cornerstone of quantum theory that reshapes our understanding of the fundamental principles of physics.

Implications of Uncertainty

The implications of Heisenberg's Uncertainty Principle are profound, touching upon various aspects of quantum mechanics, measurement theory, and our understanding of the fundamental nature of reality. In this section, we explore some of the key implications of uncertainty in the quantum realm.

1. Limitations of Measurement:

Heisenberg's Uncertainty Principle imposes fundamental limits on our ability to simultaneously measure certain pairs of physical properties, such as position and momentum, with arbitrary precision. This implies that there are inherent limitations to our knowledge of the quantum world, as we cannot precisely determine both the position and momentum of a particle simultaneously.

2. Wave-Particle Duality:

The uncertainty principle is intimately related to the wave-particle duality of matter and radiation. It underscores the dual nature of particles, which can exhibit both wave-like and particle-like behavior depending on the context of the experiment. The uncertainty in position and momentum arises from the wave-like nature of particles, where their properties are spread out over a range of possible states.

3. Quantum Superposition:

Uncertainty is inherent in the concept of quantum superposition, where particles can exist in multiple states simultaneously until measured or observed. The uncertainty principle quantifies the limits of our knowledge about the state of a particle in superposition, as we cannot precisely determine both its position and momentum at the same time.

4. Observer Effect:

The uncertainty principle highlights the role of observation and measurement in quantum mechanics. The act of measuring a quantum system inevitably disturbs its state, leading to uncertainty in subsequent measurements. This phenomenon, known as the observer effect, underscores the intimate connection between the observer and the observed system in quantum mechanics.

5. Quantum Mechanics and Technology:

The uncertainty principle has practical implications for the development of quantum technologies, such as quantum computing and quantum cryptography. It places constraints on the precision with which quantum systems can be manipulated and measured, influencing the design and operation of quantum devices.

6. Philosophical Implications:

Heisenberg's Uncertainty Principle has profound philosophical implications, challenging our classical intuition about the nature of reality. It suggests that there are inherent limitations to our knowledge of the universe and raises questions about the nature of measurement, observation, and the fundamental nature of reality at the quantum level.

In summary, the implications of Heisenberg's Uncertainty Principle extend far beyond the realm of physics, touching upon fundamental questions about the nature of reality, the limits of knowledge, and the role of observation in shaping our understanding of the quantum world. As we continue to explore the mysteries of quantum mechanics, uncertainty remains a central concept that reshapes our perspective on the fundamental principles of the universe.

Measurement Problem and Quantum Decoherence

The measurement problem and quantum decoherence are two interconnected concepts in quantum mechanics that address the challenges associated with measurement and the emergence of classical behavior from quantum systems. In this section, we explore both the measurement problem and quantum decoherence and their implications for our understanding of the quantum world.

Measurement Problem:

The measurement problem arises from the fundamental principles of quantum mechanics and concerns the nature of measurement in quantum systems. According to quantum theory, particles exist in a state of superposition, where they can simultaneously occupy multiple states until measured or observed. However, when a measurement is made, the wave function of the particle collapses to a single, well-defined state, leading to the emergence of classical behavior. The measurement problem raises several philosophical and interpretational questions about the nature of measurement and the role of the observer in quantum mechanics. It challenges our classical intuition about the determinacy of physical systems and highlights the mysterious and inherently probabilistic nature of quantum phenomena.

Quantum Decoherence:

Quantum decoherence is a physical process that explains how classical behavior emerges from quantum systems as they interact with their environment. When a quantum system interacts with its surrounding environment, such as through collisions with air molecules or interactions with electromagnetic radiation, the coherence of the system's quantum state is gradually lost. As a result of decoherence, the quantum system becomes entangled with its environment, leading to the suppression of interference effects and the emergence of classical behavior. Decoherence explains why macroscopic objects, such as everyday objects, appear to exhibit classical behavior despite being composed of quantum particles.

Connection between Measurement Problem and Decoherence:

Quantum decoherence provides a resolution to the measurement problem by explaining how classical behavior arises from quantum systems without invoking the need for conscious observation. Decoherence occurs naturally as quantum systems interact with their environment, leading to the suppression of quantum interference effects and the emergence of classical-like behavior.

While decoherence resolves the measurement problem by providing a physical mechanism for the collapse of the wave function, it also raises questions about the interpretation of quantum mechanics and the nature of reality. Some interpretations, such as the many-worlds interpretation, suggest that decoherence leads to the branching of multiple parallel universes, each corresponding to a different outcome of a quantum measurement.

Implications and Future Directions:

The measurement problem and quantum decoherence have far-reaching implications for our understanding of the quantum world and the development of quantum technologies. They highlight the intricate interplay between quantum and classical behavior and the challenges associated with measurement and observation at the quantum level.

Further research into the measurement problem and quantum decoherence may lead to new insights into the nature of reality and the development of novel quantum technologies, such as quantum computers and quantum communication systems. These concepts continue to inspire ongoing debate and exploration at the forefront of modern physics.

Chapter 4: Quantum Mechanics Basics

Quantum mechanics forms the foundation of modern physics, providing a framework for understanding the behavior of particles at the atomic and subatomic levels. In this chapter, we explore the fundamental principles and concepts of quantum mechanics, laying the groundwork for a deeper understanding of the quantum world.

Historical Background

We begin with a brief overview of the historical development of quantum mechanics, tracing its origins from the early 20th century to the present day. Key milestones, such as the formulation of Planck's quantum hypothesis, Einstein's explanation of the photoelectric effect, and the development of wave mechanics and matrix mechanics, are discussed.

Principles of Quantum Mechanics

We then delve into the core principles of quantum mechanics, including:
1. Wave-Particle Duality: The concept that particles, such as electrons and photons, exhibit both wave-like and particle-like behavior.
2. Superposition: The idea that particles can exist in multiple states simultaneously until measured or observed.
3. Uncertainty Principle: The principle formulated by Heisenberg, which states that there is a fundamental limit to the precision with which certain pairs of properties, such as position and momentum, can be simultaneously known.
4. Wave Function and Probability Amplitude: The mathematical description of quantum states, represented by wave functions, which encode the probability amplitudes of particles in different states.

Mathematical Formalism

We introduce the mathematical formalism of quantum mechanics, including:
1. Schrödinger's Equation: The fundamental equation that describes the time evolution of quantum systems and predicts the behavior of particles in various states.
2. Operators and Observables: The mathematical operators that represent physical observables, such as position, momentum, and energy, and their corresponding eigenvalues.
3. Wave Function Collapse: The phenomenon where the wave function of a particle collapses to a single state upon measurement, leading to a definite outcome.

Interpretations of Quantum Mechanics

We explore different interpretations of quantum mechanics, including the Copenhagen interpretation, the many-worlds interpretation, and the pilot-wave theory. Each interpretation offers a different perspective on the nature of reality and the role of measurement in quantum mechanics.

Applications of Quantum Mechanics

Finally, we discuss some of the practical applications of quantum mechanics in various fields, such as:
1. Quantum Computing: The use of quantum mechanical phenomena, such as superposition and entanglement, to perform computations more efficiently than classical computers.

2. Quantum Cryptography: The development of secure communication protocols based on the principles of quantum mechanics, such as quantum key distribution.
3. Quantum Sensing and Metrology: The use of quantum systems for high-precision measurements, such as atomic clocks and quantum sensors.

Conclusion

In conclusion, quantum mechanics provides a rich and intricate framework for understanding the behavior of particles at the quantum level. By exploring its fundamental principles, mathematical formalism, interpretations, and applications, we gain a deeper appreciation for the profound implications of quantum theory and its role in shaping our understanding of the universe.

Wave Function and Probability Amplitude

Wave Function and Probability Amplitude
In quantum mechanics, the wave function is a fundamental concept that describes the quantum state of a system, such as a particle or a collection of particles. The wave function encodes information about the probability distribution of possible outcomes of measurements that can be made on the system. Additionally, the probability amplitude, derived from the wave function, provides insight into the likelihood of finding a particle in a particular state.

Wave Function:

The wave function, typically denoted by the symbol Ψ (psi), is a mathematical function that depends on the spatial coordinates and time. It describes the state of a quantum system and contains information about the possible positions, momenta, energies, and other properties of the system.
Mathematically, the wave function $\Psi(x, y, z, t)$ represents the amplitude of the probability wave associated with finding a particle at a given position (x, y, z) and time t. The square of the absolute value of the wave function, $|\Psi(x, y, z, t)|^2$, gives the probability density of finding the particle at a specific position.

Probability Amplitude:

The probability amplitude is a complex number derived from the wave function that quantifies the likelihood of finding a particle in a particular state. Unlike classical probability, which is represented by real numbers between 0 and 1, probability amplitudes can be complex numbers with both magnitude and phase.
The probability amplitude for a particular state corresponds to the coefficient of the wave function associated with that state. When multiple states are superimposed, the probability amplitudes of each state interfere with each other, leading to interference effects characteristic of quantum mechanics.

Interpretation:

In the Copenhagen interpretation of quantum mechanics, the square of the absolute value of the probability amplitude, |A|^2, gives the probability of observing a particular measurement outcome corresponding to that state. This interpretation highlights the probabilistic nature of quantum mechanics, where measurements yield probabilistic outcomes rather than deterministic ones.

Applications:

The wave function and probability amplitudes play a crucial role in many aspects of quantum mechanics, including the prediction of experimental outcomes, the formulation of quantum algorithms in quantum computing, and the understanding of phenomena such as quantum entanglement and quantum tunneling.

Conclusion:

The wave function and probability amplitudes are foundational concepts in quantum mechanics, providing a mathematical framework for describing the behavior of particles at the quantum level. By understanding these concepts, physicists can make predictions about the behavior of quantum systems and unlock the potential of quantum technologies for various applications.

Schrödinger Equation

The Schrödinger equation is a fundamental equation in quantum mechanics that describes how the wave function of a quantum system evolves over time. It provides a mathematical framework for understanding the behavior of particles at the atomic and subatomic levels and plays a central role in the prediction of experimental outcomes and the development of quantum technologies. In this section, we explore the Schrödinger equation and its significance in quantum mechanics.

Formulation:

The time-dependent Schrödinger equation, first formulated by Austrian physicist Erwin Schrödinger in 1925, describes the evolution of the wave function $\Psi(x, t)$ of a quantum system with respect to time. It is given by:

$i\hbar \partial \Psi \partial t = H^\wedge \Psi$

$i\hbar$

∂t

$\partial \Psi$

$=$

H

$\hat{\Psi}$

Where:
- i
- i is the imaginary unit,
- $\hbar$
- $\hbar$ is the reduced Planck constant (
- $\hbar/(2\pi)$
- h/(2π)),
- $\partial\Psi \partial t$
- ∂t
- $\partial\Psi$
-
- is the partial derivative of the wave function with respect to time,
- $\hat{H}$
- H
- ^
- is the Hamiltonian operator, representing the total energy of the system.

Interpretation:

The Schrödinger equation embodies the wave nature of particles in quantum mechanics. It describes how the wave function evolves over time under the influence of the Hamiltonian operator, which encodes the energy of the system. Solutions to the Schrödinger equation provide information about the possible states of the system and the probabilities of different measurement outcomes.

Time-Independent Schrödinger Equation:

In certain cases, the Hamiltonian operator may not depend explicitly on time, leading to a simplified version of the Schrödinger equation known as the time-independent Schrödinger equation:

$\hat{H}\Psi = E\Psi$
H
^
$\Psi = E\Psi$

Where:
- $\hat{H}$
- H
- ^
- is the Hamiltonian operator,
- Ψ
- Ψ is the wave function,
- E
- E is the energy of the system.

The time-independent Schrödinger equation is often used to determine the energy eigenstates and eigenvalues of a quantum system, providing insights into its quantized energy levels.

Solutions and Applications:

Solutions to the Schrödinger equation yield the wave function of the quantum system, which contains information about the probability distribution of possible measurement outcomes. The Schrödinger equation has broad applications in quantum mechanics, including the prediction of atomic and molecular spectra, the behavior of quantum systems in external fields, and the development of quantum algorithms in quantum computing.

Conclusion:

The Schrödinger equation stands as a cornerstone of quantum mechanics, providing a powerful mathematical framework for understanding the behavior of particles at the quantum level. By solving the Schrödinger equation, physicists can make predictions about the behavior of quantum systems and unlock the potential of quantum technologies for various applications, from cryptography to materials science.

Operators and Observables

In quantum mechanics, operators and observables are fundamental concepts that play a central role in describing the physical properties of quantum systems and making predictions about measurement outcomes. In this section, we explore the relationship between operators and observables and their significance in quantum mechanics.

Operators:

Operators in quantum mechanics are mathematical objects that act on wave functions to produce new wave functions or numerical values. These operators represent physical observables, such as position, momentum, energy, and angular momentum, that can be measured in experiments.

Examples of Operators:

1. Position Operator (
2. $x^\wedge$
3. x
4. ^
5.): This operator represents the position of a particle along a particular axis in space.
6. Momentum Operator (
7. $p^\wedge$
8. p

9. ^
10.
11.): This operator represents the momentum of a particle along a particular axis in space.
12. Hamiltonian Operator (
13. $H^\wedge$
14. H
15. ^
16.): This operator represents the total energy of a quantum system, including kinetic and potential energy contributions.

Observables:

Observables in quantum mechanics are physical properties of a system that can be measured experimentally. These properties correspond to eigenvalues of the corresponding operators in the mathematical formalism of quantum mechanics.

Examples of Observables:

1. Position (
2. x
3. x): The observable representing the spatial location of a particle.
4. Momentum (
5. p
6. p): The observable representing the linear momentum of a particle.
7. Energy (
8. E
9. E): The observable representing the total energy of a quantum system.

Relationship between Operators and Observables:

In quantum mechanics, the connection between operators and observables is established through the eigenvalue equation. For an observable represented by an operator
$A^\wedge$
A
^
, the eigenvalue equation is given by:
$A^\wedge|\psi\rangle = a|\psi\rangle$
A
^
$|\psi\rangle = a|\psi\rangle$
Where:
- $A^\wedge$
- A
- ^
- is the operator representing the observable,
- $|\psi\rangle$

- $|\psi\rangle$ is the wave function of the quantum system,
- a
- a is the eigenvalue corresponding to the observable.

The eigenvalue equation states that when the operator acts on the wave function, it yields the wave function multiplied by a scalar factor, which represents the measurement outcome (eigenvalue) of the observable.

Measurement Process:

During a measurement in quantum mechanics, an observable is measured, and one of its eigenvalues is obtained as the measurement outcome. The measurement process causes the wave function of the system to collapse to the eigenstate corresponding to the measured eigenvalue.

Conclusion:

Operators and observables are essential concepts in quantum mechanics, providing a mathematical framework for describing the physical properties of quantum systems and making predictions about measurement outcomes. By understanding the relationship between operators and observables, physicists can analyze the behavior of quantum systems and interpret experimental results in the context of quantum mechanics.

Chapter 5: Quantum States and Quantum Entanglement

In this chapter, we delve into the fascinating world of quantum states and quantum entanglement, two phenomena that lie at the heart of quantum mechanics and have profound implications for our understanding of the quantum world.

Quantum States:

Quantum states describe the complete set of physical properties of a quantum system, including its position, momentum, energy, and other observable quantities. In quantum mechanics, these properties are represented by wave functions, which encode information about the probabilities of different measurement outcomes. We explore the mathematical formalism of quantum states and their significance in predicting experimental results.

Superposition:

One of the defining features of quantum states is superposition, where particles can exist in multiple states simultaneously until measured or observed. We discuss how superposition leads to interference phenomena, such as those observed in the double-slit experiment, and how it underscores the probabilistic nature of quantum mechanics.

Quantum Entanglement:

Quantum entanglement is a phenomenon in which the quantum states of two or more particles become correlated in such a way that the state of one particle cannot be described independently of the state of the others. We explore the concept of entanglement and its implications for quantum information processing, quantum teleportation, and the foundations of quantum mechanics.

Bell's Theorem:

Bell's theorem, proposed by physicist John Bell in 1964, provides a way to test the validity of quantum mechanics against classical theories with local hidden variables. Experimental tests of Bell's theorem have confirmed the existence of quantum entanglement and ruled out certain types of classical explanations for quantum correlations, providing strong evidence for the non-local nature of quantum mechanics.

Applications of Quantum Entanglement:

Quantum entanglement has practical applications in various fields, including quantum cryptography, quantum teleportation, and quantum computing. We discuss how entanglement enables secure communication protocols, allows for the teleportation of quantum states between distant locations, and forms the basis of quantum algorithms in quantum computing.

Entanglement and Spooky Action at a Distance:

Entanglement has been described by Einstein as "spooky action at a distance" due to its non-local nature, where the state of one particle can instantaneously influence the state of another, even when separated by large distances. We explore the philosophical implications of entanglement and its implications for our understanding of the fundamental nature of reality.

Conclusion:

Quantum states and quantum entanglement are fascinating phenomena that challenge our classical intuition about the nature of reality. By exploring these concepts, we gain a deeper understanding of the rich and complex world of quantum mechanics and the profound implications for technology, communication, and our understanding of the universe.

Quantum States: Pure vs. Mixed

In quantum mechanics, quantum states can be classified into two main categories: pure states and mixed states. These two types of states have distinct characteristics and play different roles in describing the behavior of quantum systems. In this section, we explore the differences between pure and mixed quantum states and their significance in quantum mechanics.

Pure States:

Pure states represent the most idealized form of quantum states, where the quantum system is described by a single, definite wave function. Mathematically, a pure state is represented by a normalized wave function,
$|\psi\rangle$
$|\psi\rangle$, which contains all the information about the quantum system and corresponds to a specific quantum state with certainty.

Characteristics of Pure States:
1. Definiteness: Pure states are characterized by a well-defined quantum state, with all observable properties of the system determined by the wave function.
2. Superposition: Pure states can exhibit superposition, where the system exists in a linear combination of multiple states simultaneously until measured or observed.

Mixed States:

Mixed states represent a statistical ensemble of pure states, where the quantum system is in a probabilistic mixture of different states. Unlike pure states, mixed states do not have a single, definite wave function but instead are described by a density operator,
$\hat{\varrho}$
$\hat{\rho}$

, which accounts for the probabilities of different pure states in the ensemble.

Characteristics of Mixed States:
1. Probabilistic Nature: Mixed states arise from uncertainty about the true quantum state of the system, leading to probabilistic outcomes in measurements.
2. Mixedness: Mixed states exhibit a degree of "mixedness," reflecting the statistical mixture of different pure states in the ensemble.

Relation between Pure and Mixed States:

Mixed states can be viewed as a statistical mixture of pure states, where each pure state contributes to the overall mixture with a certain probability weight. Conversely, a pure state can be considered a special case of a mixed state, where the density operator reduces to a projector onto the pure state.

Significance in Quantum Mechanics:

Pure states are often used to describe idealized quantum systems and serve as the basis for theoretical calculations and predictions. Mixed states, on the other hand, are encountered in situations involving uncertainty, noise, or incomplete information about the quantum state of the system. Understanding the distinction between pure and mixed states is crucial for interpreting experimental results and analyzing the behavior of quantum systems in practical applications.

Conclusion:

In summary, pure and mixed states are two fundamental types of quantum states that describe the behavior of quantum systems in different contexts. While pure states represent well-defined quantum states with certainty, mixed states arise from uncertainty and statistical mixtures of different pure states. By understanding the distinction between pure and mixed states, physicists can gain insights into the probabilistic nature of quantum mechanics and its implications for real-world applications.

Entanglement: Spooky Action at a Distance

Quantum entanglement is a phenomenon in quantum mechanics where the quantum states of two or more particles become correlated in such a way that the state of one particle cannot be described independently of the state of the others, regardless of the distance between them. This concept was famously referred to by Albert Einstein as "spooky action at a distance." In this section, we delve into the fascinating phenomenon of entanglement and its implications for our understanding of the quantum world.

The Concept of Entanglement:

Entanglement arises when two or more particles are generated or interact in such a way that their quantum states become inseparably linked, forming a single entangled system. This means that the properties of one particle are intrinsically connected to the properties of the other particles, even when they are separated by vast distances.

EPR Paradox and Einstein's Skepticism:

The concept of entanglement was famously discussed in a paper by Einstein, Podolsky, and Rosen (EPR) in 1935, where they highlighted what they saw as a "paradox" in quantum mechanics. They argued that entanglement implied an instantaneous connection between particles separated by large distances, violating the principle of locality and suggesting "spooky action at a distance."

Bell's Theorem and Experimental Tests:

In 1964, physicist John Bell formulated a theorem that provided a way to test the validity of quantum mechanics against classical theories with local hidden variables. Experimental tests of Bell's theorem have since confirmed the existence of quantum entanglement and ruled out certain types of classical explanations for quantum correlations.

Non-Locality and Quantum Information:

Entanglement implies a form of non-locality, where the state of one particle can instantaneously influence the state of another, regardless of the distance between them. This non-local correlation has profound implications for quantum information processing, quantum teleportation, and the security of quantum cryptography protocols.

Interpretations and Philosophical Implications:

The phenomenon of entanglement raises deep questions about the nature of reality, the role of observation in quantum mechanics, and the fundamental principles of physics. It challenges our classical intuition about causality and locality, suggesting that the quantum world operates according to rules that are fundamentally different from classical physics.

Conclusion:

Entanglement remains one of the most intriguing and mysterious phenomena in quantum mechanics, with profound implications for our understanding of the quantum world and the development of quantum technologies. By exploring the concept of entanglement and its implications, physicists continue to push the boundaries of our knowledge and uncover the secrets of the quantum universe.

Bell's Theorem and Tests of Quantum Entanglement

Bell's theorem, formulated by physicist John Bell in 1964, provides a way to test the validity of quantum mechanics against classical theories with local hidden variables. The theorem addresses the EPR paradox (Einstein-Podolsky-Rosen paradox) and the concept of entanglement, which Einstein famously referred to as "spooky action at a distance." In this section, we explore Bell's theorem and its significance in testing the principles of quantum mechanics through experimental tests of quantum entanglement.

Bell's Inequality:

Bell's theorem starts with the derivation of Bell's inequality, which is a mathematical inequality that constrains the correlations between measurements made on entangled particles under certain assumptions. Bell showed that if the predictions of quantum mechanics are correct, these correlations should violate Bell's inequality, whereas classical theories with local hidden variables should satisfy it.

Experimental Setup:

Experimental tests of Bell's theorem typically involve the creation of entangled particle pairs, such as photons or ions, and the measurement of certain properties, such as polarization or spin, on each particle. The measurements are made at distant locations, often separated by large distances, to test for non-local correlations between the particles.

Violation of Bell's Inequality:

Experimental tests of Bell's theorem have consistently shown that the correlations between entangled particles violate Bell's inequality, in accordance with the predictions of quantum mechanics. This implies that the correlations between the particles cannot be explained by classical theories with local hidden variables and provides strong evidence for the reality of quantum entanglement.

Aspect Experiments:

One of the pioneering experiments testing Bell's theorem was conducted by Alain Aspect and his colleagues in the 1980s. The Aspect experiments involved measuring the polarization of entangled photon pairs and demonstrated violations of Bell's inequality, confirming the existence of quantum entanglement and ruling out certain types of classical explanations for the observed correlations.

Tests of Quantum Entanglement:

Experimental tests of Bell's theorem continue to be conducted using increasingly sophisticated setups and improved measurement techniques. These tests not only confirm the predictions of quantum mechanics but also provide insights into the nature of entanglement and its potential applications in quantum information processing, quantum cryptography, and quantum communication.

Conclusion:

Bell's theorem and experimental tests of quantum entanglement have revolutionized our understanding of the quantum world and provided experimental validation of the principles of quantum mechanics. By challenging classical notions of locality and causality, these tests have opened up new avenues for research and technology development in the field of quantum information science, paving the way for a future of quantum-enabled technologies and applications.

Chapter 6: Quantum Mechanics in Practice

Quantum mechanics has far-reaching implications beyond theoretical physics, with practical applications in various fields ranging from technology and engineering to cryptography and medicine. In this chapter, we explore how quantum mechanics is applied in practice, driving innovation and revolutionizing industries.

Quantum Computing:

Quantum computing harnesses the principles of quantum mechanics to perform computations that are beyond the capabilities of classical computers. We discuss the development of quantum algorithms, quantum gates, and quantum processors, as well as the potential applications of quantum computing in optimization, cryptography, and simulation of quantum systems.

Quantum Cryptography:

Quantum cryptography leverages the principles of quantum mechanics to secure communication channels against eavesdropping and hacking. We explore quantum key distribution protocols, such as BB84 and E91, which rely on the properties of quantum states, such as superposition and entanglement, to achieve unconditional security.

Quantum Sensing and Metrology:

Quantum sensing and metrology utilize quantum systems to make highly precise measurements of physical quantities, such as time, frequency, and magnetic field. We discuss the development of atomic clocks, magnetometers, and gravimeters based on principles of quantum mechanics, as well as their applications in navigation, geophysics, and fundamental research.

Quantum Communication:

Quantum communication enables secure and efficient transmission of information using quantum states as carriers of information. We explore the principles of quantum teleportation and quantum teleportation-based protocols for secure communication, as well as the development of quantum communication networks and quantum internet infrastructure.

Quantum Materials and Nanotechnology:

Quantum mechanics plays a crucial role in understanding the behavior of materials at the nanoscale and designing novel materials with unique properties. We discuss the development of quantum dots, graphene, and other quantum materials, as well as their applications in electronics, photonics, and energy storage.

Quantum Biology and Medicine:

Quantum mechanics is increasingly being applied to study biological systems and develop new medical technologies. We explore the role of quantum mechanics in understanding biological processes, such as photosynthesis and enzyme catalysis, as well as the development of quantum-inspired medical imaging techniques and drug delivery systems.

Conclusion:

Quantum mechanics is not just a theoretical framework but a powerful tool for innovation and discovery across a wide range of disciplines. By harnessing the principles of quantum mechanics, researchers and engineers are pushing the boundaries of what is possible, driving progress and shaping the future of technology, science, and society.

Quantum Tunneling

Quantum tunneling is a fascinating phenomenon in quantum mechanics where particles penetrate through potential energy barriers that would be classically impassable. In this section, we explore the concept of quantum tunneling, its implications, and its applications in various fields.

The Tunneling Effect:

In classical physics, particles are confined to regions of space where their kinetic energy exceeds the potential energy of the barriers they encounter. However, in quantum mechanics, particles exhibit wave-like behavior and can tunnel through barriers even when their energy is lower than the barrier height.

Understanding Tunneling:

According to quantum mechanics, particles are described by wave functions that extend over space, allowing them to penetrate into classically forbidden regions. This phenomenon arises from the wave-particle duality and the probabilistic nature of quantum mechanics, where particles have a finite probability of tunneling through barriers.

Mathematical Description:

The probability of tunneling through a potential energy barrier is described by the wave function of the particle and depends on various factors, including the width and height of the barrier, the mass and energy of the particle, and the angle of incidence. Quantum tunneling is typically described using mathematical formalisms such as the Schrödinger equation and tunneling probability calculations.

Implications and Applications:

1. Nuclear Fusion: Quantum tunneling plays a crucial role in nuclear fusion reactions, where atomic nuclei overcome the Coulomb barrier to fuse and release energy.
2. Semiconductor Devices: In semiconductor devices such as tunnel diodes and flash memory, tunneling currents are utilized for electronic applications.
3. Scanning Tunneling Microscopy: Quantum tunneling is the underlying principle of scanning tunneling microscopy (STM), a powerful technique used to image surfaces at the atomic level.
4. Nanotechnology: Tunneling phenomena are exploited in nanotechnology for the fabrication of nanoscale devices, quantum dots, and tunnel junctions.
5. Quantum Computing: Tunneling is a key mechanism in quantum computing, where qubits tunnel between different states to perform quantum operations.

Experimental Observations:

Quantum tunneling has been observed and confirmed in numerous experimental studies across various fields, providing strong evidence for the validity of quantum mechanics. These experiments have verified the predictions of tunneling theory and demonstrated its practical implications.

Conclusion:

Quantum tunneling is a remarkable manifestation of the principles of quantum mechanics, allowing particles to penetrate through barriers that would be classically impenetrable. By understanding and harnessing the phenomenon of tunneling, scientists and engineers are developing new technologies and advancing our understanding of the quantum world.

Quantum Computing

Quantum computing is a revolutionary paradigm of computation that harnesses the principles of quantum mechanics to perform calculations that would be infeasible for classical computers. In this section, we delve into the principles of quantum computing, its potential applications, and the challenges and advancements in the field.

Principles of Quantum Computing:

Quantum computing leverages two fundamental principles of quantum mechanics: superposition and entanglement. Quantum bits, or qubits, can exist in superposition states, representing both 0 and 1 simultaneously. Additionally, qubits can be entangled, such that the state of one qubit is dependent on the state of another, even when they are separated by large distances.

Quantum Gates and Algorithms:

Quantum computations are performed using quantum gates, which manipulate the quantum states of qubits to perform operations. Quantum algorithms, such as Shor's algorithm for integer factorization and Grover's algorithm for database search, exploit the parallelism and interference effects of quantum systems to solve certain problems exponentially faster than classical algorithms.

Types of Quantum Computers:

1. Universal Quantum Computers: These are general-purpose quantum computers capable of executing a wide range of quantum algorithms. Universal quantum computers are still in the early stages of development but hold the promise of revolutionizing various fields, including cryptography, materials science, and optimization.
2. Special-Purpose Quantum Computers: These are quantum computers optimized for specific tasks, such as optimization problems, simulation of quantum systems, or cryptographic applications. Special-purpose quantum computers may offer more efficient solutions to certain classes of problems compared to classical approaches.

Applications of Quantum Computing:

1. Cryptography: Quantum computers have the potential to break classical cryptographic protocols, such as RSA and ECC, by efficiently factoring large numbers or solving the discrete logarithm problem. Quantum-safe cryptographic algorithms, based on quantum-resistant primitives, are being developed to mitigate these threats.
2. Optimization and Machine Learning: Quantum computers can tackle optimization problems, such as portfolio optimization, logistics planning, and drug discovery, more efficiently than classical algorithms. Quantum machine learning algorithms, such as quantum neural networks and quantum support vector machines, offer potential improvements in pattern recognition and data analysis.
3. Materials Science and Simulation: Quantum computers enable the simulation of quantum systems, such as chemical reactions, materials properties, and quantum many-body systems, with unprecedented accuracy and efficiency. This has implications for the discovery of new materials, drugs, and catalysts.

Challenges and Advancements:

Despite the promise of quantum computing, significant challenges remain, including qubit coherence and error correction, scalability, and fault tolerance. Researchers are making strides in developing error-correction codes, improving qubit coherence times, and scaling up quantum systems to overcome these challenges and realize the full potential of quantum computing.

Conclusion:

Quantum computing represents a paradigm shift in computation, offering the potential to solve complex problems that are beyond the reach of classical computers. By harnessing the principles of quantum mechanics, researchers are paving the way for a new era of computing and unlocking unprecedented possibilities in science, technology, and innovation.

Quantum Cryptography

Quantum cryptography is a field of cryptography that leverages the principles of quantum mechanics to provide secure communication protocols that are theoretically immune to eavesdropping and hacking. In this section, we explore the principles of quantum cryptography, its applications, and its implications for secure communication.

Principles of Quantum Cryptography:

Quantum cryptography relies on the fundamental principles of quantum mechanics, including superposition, entanglement, and the Heisenberg uncertainty principle. Quantum key distribution (QKD) protocols, such as BB84 and E91, exploit these principles to establish secure cryptographic keys between two parties.

Quantum Key Distribution (QKD):

In a quantum key distribution protocol, two parties, traditionally referred to as Alice and Bob, exchange quantum states encoded with cryptographic information. These quantum states are transmitted over a quantum communication channel, such as a fiber optic cable or free-space optical link, and are susceptible to interception.

BB84 Protocol:

The BB84 protocol, proposed by Charles Bennett and Gilles Brassard in 1984, is one of the earliest and most widely studied QKD protocols. In the BB84 protocol, Alice prepares qubits in one of two orthogonal bases (e.g., the Z basis and the X basis) and sends them to Bob. Bob randomly chooses a basis to measure each qubit and communicates his measurement basis to Alice. Alice and Bob discard incompatible measurements and use the remaining qubits to generate a shared secret key.

E91 Protocol:

The E91 protocol, proposed by Artur Ekert in 1991, is based on the phenomenon of quantum entanglement. In the E91 protocol, Alice and Bob share entangled pairs of particles (e.g., photons) generated using a source of entanglement. By performing measurements on their respective particles, Alice and Bob can generate a shared secret key with enhanced security guarantees.

Security of Quantum Cryptography:

The security of quantum cryptography protocols is rooted in the principles of quantum mechanics, which prevent an eavesdropper, traditionally referred to as Eve, from intercepting or measuring the quantum states without being detected. Any attempt by Eve to intercept the quantum states would disturb the quantum states, leading to detectable errors in the communication.

Applications of Quantum Cryptography:

1. Secure Communication: Quantum cryptography provides a means of establishing secure communication channels between parties, ensuring the confidentiality and integrity of transmitted data.
2. Quantum Key Distribution Networks: Quantum cryptography can be deployed in quantum key distribution networks, allowing multiple parties to securely exchange cryptographic keys over long distances.
3. Quantum-Safe Cryptography: Quantum cryptography also has implications for post-quantum cryptography, where cryptographic algorithms are designed to be resistant to attacks by quantum computers.

Challenges and Advancements:

Despite the promise of quantum cryptography, practical implementations face challenges related to technological limitations, including qubit coherence times, transmission distances, and scalability. Researchers are actively working on developing practical quantum cryptographic systems and overcoming these challenges to realize the full potential of quantum cryptography.

Conclusion:

Quantum cryptography offers a revolutionary approach to secure communication, leveraging the principles of quantum mechanics to provide theoretically unbreakable cryptographic protocols. By harnessing the unique properties of quantum systems, quantum cryptography has the potential to revolutionize the field of cybersecurity and ensure the confidentiality and integrity of sensitive information in an increasingly interconnected world.

Chapter 7: Quantum Field Theory

Quantum Field Theory (QFT) is a theoretical framework that extends the principles of quantum mechanics to describe the behavior of particles and fields at the fundamental level. In this chapter, we explore the foundations of quantum field theory, its mathematical formalism, and its applications in particle physics and beyond.

Foundations of Quantum Field Theory:

1. Classical Field Theory: Quantum field theory builds upon classical field theory, which describes the dynamics of fields, such as the electromagnetic field, using classical equations of motion.
2. Quantization of Fields: In quantum field theory, classical fields are quantized, meaning that the field values become operators that satisfy quantum commutation relations. This leads to the creation and annihilation of particles, described by field excitations.

Mathematical Formalism:

1. Lagrangian Formalism: Quantum field theory is formulated using the Lagrangian density, which specifies the dynamics of the fields and their interactions. The Lagrangian density is invariant under certain symmetries, such as gauge symmetry, which play a crucial role in the formulation of quantum field theories.
2. Path Integral Formulation: Another approach to quantum field theory is the path integral formulation, developed by Richard Feynman. This formulation treats the evolution of fields as the sum over all possible paths in configuration space, weighted by the exponential of the action.

Quantum Electrodynamics (QED):

Quantum electrodynamics is the quantum field theory describing the electromagnetic interaction between charged particles. QED successfully combines quantum mechanics with special relativity and accurately describes phenomena such as the Lamb shift, electron magnetic moment, and electromagnetic scattering processes.

Standard Model of Particle Physics:

The Standard Model is a quantum field theory that describes the fundamental particles and their interactions via three fundamental forces: electromagnetism, weak nuclear force, and strong nuclear force. The Standard Model encompasses quantum electrodynamics (QED), electroweak theory, and quantum chromodynamics (QCD) and has been extensively tested through experiments at particle accelerators.

Quantum Field Theory in Cosmology and Astrophysics:

Quantum field theory plays a significant role in understanding the early universe, black holes, and other astrophysical phenomena. In cosmology, quantum field theory is used to study inflationary models, the cosmic microwave background radiation, and the formation of large-scale structure in the universe.

Open Problems and Future Directions:

Despite its successes, quantum field theory faces several challenges, including the quantization of gravity, the hierarchy problem, and the nature of dark matter and dark energy. Future research directions include the development of quantum field theories beyond the Standard Model and the exploration of novel phenomena in high-energy physics and cosmology.

Conclusion:

Quantum field theory stands as one of the most successful and comprehensive frameworks in theoretical physics, providing a unified description of particles and fields at the quantum level. By combining the principles of quantum mechanics with the principles of special relativity and symmetry, quantum field theory has revolutionized our understanding of the fundamental forces and particles of nature, paving the way for new discoveries and breakthroughs in particle physics, cosmology, and beyond.

Introduction to Quantum Field Theory

Quantum Field Theory (QFT) is a powerful theoretical framework that unifies the principles of quantum mechanics with the concepts of fields from classical physics. It provides a comprehensive description of fundamental particles and their interactions, offering insights into the behavior of matter and forces at the smallest scales of the universe. In this introductory section, we explore the key concepts and foundations of quantum field theory.

Classical Field Theory:

Classical field theory describes physical phenomena in terms of continuous fields that permeate spacetime. Examples include the electromagnetic field, described by Maxwell's equations, and the gravitational field, described by Einstein's theory of general relativity. In classical field theory, fields evolve according to deterministic equations of motion.

Quantum Mechanics and Particle-Wave Duality:

Quantum mechanics revolutionized our understanding of nature by introducing the concept of wave-particle duality. According to quantum mechanics, particles exhibit both particle-like and wave-like behavior. The wave function, described by Schrödinger's equation, encodes the probability amplitude of finding a particle at a given position and time.

Extension to Quantum Fields:

In quantum field theory, the fields of classical physics are quantized, meaning that field values become operators that act on a quantum state. These operators create and annihilate particles, which are excitations of the underlying quantum fields. The quantization process leads to the formulation of creation and annihilation operators for each type of particle.

Lagrangian Formalism:

Quantum field theory is formulated using the Lagrangian density, which encodes the dynamics of the fields and their interactions. The Lagrangian density is invariant under certain symmetries, such as gauge symmetry and Lorentz symmetry, which play a fundamental role in the formulation of quantum field theories. Variational principles are used to derive the equations of motion from the Lagrangian.

Feynman Diagrams and Perturbation Theory:

Feynman diagrams provide a graphical representation of particle interactions in quantum field theory. They encode the amplitudes for scattering processes and other physical phenomena. Perturbation theory is used to calculate these amplitudes systematically, treating interactions as small perturbations to free particle states.

Applications of Quantum Field Theory:

Quantum field theory has profound implications for our understanding of the universe and has found applications in diverse areas of physics, including particle physics, condensed matter physics, cosmology, and quantum information theory. It forms the theoretical basis for the Standard Model of particle physics, which describes the fundamental particles and their interactions.

Conclusion:

Quantum field theory represents a synthesis of quantum mechanics, special relativity, and classical field theory, providing a powerful framework for understanding the fundamental constituents of nature and their interactions. By combining the principles of quantum mechanics with the concept of fields, quantum field theory has revolutionized our understanding of particle physics and the fundamental forces of the universe, shaping our view of the cosmos at the most fundamental level.

Quantization of Fields

Quantization of fields is a fundamental concept in quantum field theory (QFT) where classical fields are transformed into quantum fields, allowing for the description of particles as excitations of these fields. In this section, we explore the process of quantizing fields and its implications in describing the behavior of particles at the quantum level.

Classical Fields:

In classical physics, fields are continuous quantities that permeate spacetime and describe physical phenomena, such as the electromagnetic field or the gravitational field. Classical fields are described by classical equations of motion, such as Maxwell's equations for electromagnetism or the Einstein field equations for gravity.

Quantization Process:

Quantization involves promoting classical fields to operators that act on quantum states, allowing for the quantized representation of field excitations, or particles. The quantization process typically involves the following steps:
1. Promotion to Operators: Classical fields are promoted to quantum field operators, where the field value becomes an operator that acts on a quantum state.
2. Canonical Commutation Relations: The field operators satisfy canonical commutation relations (for bosonic fields) or anti-commutation relations (for fermionic fields), ensuring the correct statistics for the corresponding particles.
3. Mode Expansion: Fields are expanded in terms of creation and annihilation operators, representing the excitation and de-excitation of particles in different momentum modes.
4. Vacuum State: The vacuum state is defined as the state with no particles present, corresponding to the ground state of the quantum field.

Bosonic and Fermionic Fields:

Bosonic fields describe particles with integer spin, such as photons or phonons, and their corresponding creation and annihilation operators satisfy commutation relations. Fermionic fields describe particles with half-integer spin, such as electrons or neutrinos, and their creation and annihilation operators satisfy anti-commutation relations.

Quantization of Gauge Fields:

Gauge fields, such as the electromagnetic field or the gluon field in quantum chromodynamics (QCD), are quantized with particular care due to gauge symmetry. Gauge fixing procedures and Faddeev-Popov ghosts are introduced to ensure that physical observables are gauge invariant.

Implications for Particle Physics:

Quantization of fields allows for the description of particles as excitations of underlying quantum fields. For example, photons are excitations of the electromagnetic field, while electrons are excitations of the electron field. Interactions between particles are described by terms in the Lagrangian density that involve the quantum fields and their derivatives.

Conclusion:

Quantization of fields is a key concept in quantum field theory, providing a framework for describing particles and their interactions at the quantum level. By quantizing classical fields and representing particles as excitations of these fields, quantum field theory offers a unified description of the fundamental constituents of nature and their behavior in a consistent quantum mechanical framework.

Particle Physics and Quantum Field Theory

Particle physics is the branch of physics that studies the fundamental constituents of matter and their interactions at the smallest scales of the universe. Quantum field theory (QFT) provides the theoretical framework for understanding the behavior of particles and fields in particle physics. In this section, we explore the relationship between particle physics and quantum field theory, highlighting key concepts and applications.

Quantum Field Theory and Particle Interactions:

Quantum field theory describes particles as excitations of underlying quantum fields, with interactions between particles mediated by the exchange of virtual particles. In QFT, particles are treated as quantized excitations of the corresponding fields, such as the electron field or the photon field, and interactions are described by terms in the Lagrangian density that involve these fields and their derivatives.

Standard Model of Particle Physics:

The Standard Model is the most successful quantum field theory of particle physics, describing the fundamental particles and their interactions via three fundamental forces: electromagnetism, weak nuclear force, and strong nuclear force. The Standard Model incorporates quantum electrodynamics (QED), electroweak theory, and quantum chromodynamics (QCD) into a unified framework.

Electroweak Theory:

Electroweak theory unifies the electromagnetic force with the weak nuclear force into a single electroweak force, mediated by the exchange of W and Z bosons. The theory is based on the gauge symmetry group

$$SU(2)_L \times U(1)_Y$$

and describes the interactions between quarks, leptons, and gauge bosons.

Quantum Chromodynamics (QCD):

Quantum chromodynamics is the quantum field theory describing the strong nuclear force, which binds quarks together to form hadrons, such as protons and neutrons. QCD is based on the gauge symmetry group

$$SU(3)_C$$

and describes the interactions between quarks and gluons, the force carriers of the strong force.

Higgs Mechanism and Particle Masses:

The Higgs mechanism, proposed in the 1960s, explains how particles acquire mass through the interaction with the Higgs field. In the Standard Model, the Higgs mechanism is responsible for giving mass to the W and Z bosons, as well as to fermions via Yukawa interactions with the Higgs boson.

Beyond the Standard Model:

While the Standard Model has been incredibly successful in describing experimental data, it is not a complete theory of particle physics. There are several open questions, including the nature of dark matter, the hierarchy problem, and the quantization of gravity, that motivate the search for physics beyond the Standard Model, such as supersymmetry, extra dimensions, or grand unified theories.

Experimental Verification:

Experimental verification of quantum field theory predictions is carried out at particle accelerators, such as the Large Hadron Collider (LHC) at CERN, where high-energy collisions between particles probe the fundamental interactions predicted by QFT. Experimental discoveries, such as the Higgs boson in 2012, provide crucial tests of the predictions of QFT.

Conclusion:

Quantum field theory provides the theoretical framework for understanding the fundamental particles and forces of nature in particle physics. By treating particles as excitations of quantum fields and describing their interactions via quantum field interactions, QFT has revolutionized our understanding of the subatomic world and continues to drive experimental discoveries at the forefront of particle physics research.

Chapter 8: Quantum Electrodynamics

Quantum Electrodynamics (QED) is the quantum field theory that describes the electromagnetic interaction between charged particles, such as electrons and photons. In this chapter, we delve into the principles of QED, its mathematical formalism, and its applications in describing fundamental processes in particle physics and quantum optics.

Foundations of Quantum Electrodynamics:

1. Classical Electrodynamics: QED builds upon classical electrodynamics, which describes the behavior of electric and magnetic fields using Maxwell's equations. Classical electrodynamics provides the classical limit of QED for macroscopic systems.
2. Quantum Mechanics and Fields: QED extends quantum mechanics to describe electromagnetic phenomena at the quantum level. It treats the electromagnetic field as a quantum field and describes particles, such as electrons and photons, as excitations of this field.

Mathematical Formalism:

1. Lagrangian Density: QED is formulated using the Lagrangian density, which encodes the dynamics of the electromagnetic field and the charged fermions (e.g., electrons) and their interactions. The Lagrangian density is invariant under gauge transformations, reflecting the gauge symmetry of electromagnetism.
2. Feynman Diagrams: Feynman diagrams provide a graphical representation of particle interactions in QED. They encode the probability amplitudes for scattering processes and other physical phenomena, allowing for systematic calculations of observable quantities.

Quantum Electrodynamics in Action:

1. Scattering Processes: QED describes scattering processes, such as electron-photon scattering or electron-electron scattering, where charged particles interact via the exchange of virtual photons. Feynman diagrams are used to calculate the scattering cross-sections and probabilities.

2. Radiative Corrections: QED predicts radiative corrections to observable quantities, such as the electron's magnetic moment (g-factor), due to virtual photon loops. These corrections have been measured experimentally with remarkable precision and agree with QED predictions.

Quantum Fluctuations and Vacuum Polarization:

1. Vacuum Fluctuations: QED predicts that the vacuum is not empty but filled with virtual particle-antiparticle pairs that continuously fluctuate in and out of existence. These vacuum fluctuations lead to observable effects, such as the Lamb shift and the Casimir effect.
2. Vacuum Polarization: Charged particles in the vacuum polarize the surrounding electromagnetic field, leading to a modification of the vacuum's properties. This effect has been experimentally observed and confirmed through precision measurements.

Renormalization:

1. Divergences: QED predictions often contain divergent terms, leading to infinite quantities in perturbative calculations. Renormalization techniques are employed to remove these infinities and extract physically meaningful results.
2. Finite Predictions: Through renormalization, QED yields finite predictions for observable quantities, such as scattering cross-sections and radiative corrections, which agree with experimental measurements to unprecedented precision.

Conclusion:

Quantum Electrodynamics stands as one of the most successful theories in physics, providing a precise description of the electromagnetic interaction at the quantum level. By combining the principles of quantum mechanics with the concepts of fields and gauge symmetry, QED has enabled remarkable insights into the behavior of charged particles and photons and has been validated through experimental tests with unparalleled accuracy.

Fundamentals of Quantum Electrodynamics

Quantum Electrodynamics (QED) is a quantum field theory that describes the electromagnetic interaction between charged particles, such as electrons and photons. It is one of the most successful theories in physics, accurately describing a wide range of phenomena with extraordinary precision. In this section, we delve into the fundamentals of QED, exploring its key concepts and mathematical formalism.

Quantum Fields and Particles:

1. Quantum Fields: In QED, the electromagnetic field is treated as a quantum field, with excitations corresponding to photons, the quanta of the electromagnetic field.
2. Charged Particles: Charged particles, such as electrons, are described as excitations of their respective quantum fields. In QED, electrons are treated as Dirac fields, obeying the Dirac equation.

Lagrangian Density:

The dynamics of QED are encoded in the Lagrangian density, which describes the kinetic and potential energies of the electromagnetic field and the charged particles, as well as their interactions. The QED Lagrangian density incorporates terms for the kinetic energy of the fields, the electromagnetic field strength, and the interaction between charged particles and the electromagnetic field.

Gauge Symmetry:

QED exhibits gauge symmetry under local phase transformations of the charged fields, known as gauge transformations. This symmetry ensures that the physical predictions of QED are independent of the choice of gauge, reflecting the underlying symmetry of electromagnetism.

Feynman Rules:

Feynman diagrams provide a graphical representation of particle interactions in QED. Each vertex in a Feynman diagram represents an interaction between particles, mediated by the exchange of virtual photons. Feynman rules specify how to construct and calculate the contributions of Feynman diagrams to scattering amplitudes and other observables.

Renormalization:

QED predictions often involve divergent integrals, leading to infinite results in perturbative calculations. Renormalization techniques are employed to remove these infinities and obtain finite, physically meaningful results. By renormalizing the theory, QED yields accurate predictions that agree with experimental measurements.

Quantum Fluctuations and Vacuum Polarization:

1. Vacuum Fluctuations: In QED, the vacuum is not empty but filled with virtual particle-antiparticle pairs that continuously fluctuate in and out of existence. These vacuum fluctuations contribute to observable effects, such as the Lamb shift and the Casimir effect.
2. Vacuum Polarization: Charged particles in the vacuum polarize the surrounding electromagnetic field, leading to a modification of the vacuum's properties. This effect, known as vacuum polarization, has measurable consequences, such as the correction to the electron's magnetic moment.

Experimental Confirmation:

QED predictions have been extensively tested and confirmed through experiments, with remarkable agreement between theory and observation. Experimental measurements of phenomena such as electron-photon scattering, radiative corrections, and the Lamb shift provide strong evidence for the validity of QED.

Conclusion:

Quantum Electrodynamics provides a precise and comprehensive description of the electromagnetic interaction at the quantum level. By combining the principles of quantum mechanics with the concepts of fields and gauge symmetry, QED has revolutionized our understanding of electromagnetism and continues to be a cornerstone of modern theoretical physics.

Renormalization: Dealing with Divergences

Renormalization is a crucial technique in quantum field theory (QFT) used to handle divergent integrals that arise in perturbative calculations. In this section, we explore the concept of renormalization and its role in making sense of the infinities encountered in QFT.

Origins of Divergences:

In perturbative QFT, Feynman diagrams are used to calculate scattering amplitudes and other observables by summing over all possible particle interaction processes. However, certain loop diagrams lead to integrals that diverge when evaluated.

Ultraviolet Divergences:

Ultraviolet (UV) divergences arise from loop integrals involving high momentum or short-distance physics. These divergences are associated with the behavior of quantum fields at very small distance scales and are typically encountered in loop diagrams with high loop momenta.

Infrared Divergences:

Infrared (IR) divergences arise from loop integrals involving low momentum or long-distance physics. These divergences are associated with the behavior of quantum fields at large distance scales and are typically encountered in loop diagrams involving massless particles, such as photons or gluons.

Renormalization Procedure:

The renormalization procedure involves several steps to remove divergences and obtain finite, physically meaningful results:

1. Regularization: Divergent integrals are regulated by introducing a cutoff parameter or a regulator function that regulates the behavior of the integrals at high momenta or short distances. Common regularization techniques include dimensional regularization and momentum cutoff regularization.
2. Counterterms: Counterterms are added to the Lagrangian density to cancel out the divergences introduced by loop diagrams. These counterterms contain free parameters, known as coupling constants or mass parameters, which are adjusted to absorb the divergences.
3. Renormalization Conditions: Physical observables, such as masses and coupling strengths, are defined through experimental measurements or theoretical considerations. Renormalization conditions are imposed to relate these observables to the parameters appearing in the counterterms.
4. Finite Results: By choosing appropriate renormalization conditions and adjusting the parameters of the counterterms, divergent contributions cancel out, leading to finite results for physical observables that agree with experimental measurements.

Renormalization Schemes:

Different renormalization schemes correspond to different choices of renormalization conditions and parameterizations of the counterterms. Common renormalization schemes include the minimal subtraction (MS) scheme, the modified minimal subtraction ($\overline{MS}$) scheme, and on-shell renormalization schemes.

Physical Interpretation:

While renormalization may seem ad hoc, it has a profound physical interpretation. The divergences encountered in perturbative calculations reflect the limitations of the perturbative approach and the need to account for all possible quantum fluctuations, including high-energy and short-distance effects, which are inaccessible to perturbation theory alone.

Conclusion:

Renormalization is an essential technique in quantum field theory, allowing physicists to make sense of divergent integrals encountered in perturbative calculations. By regulating, absorbing, and adjusting divergent contributions, renormalization provides a systematic framework for obtaining finite, physically meaningful results that agree with experimental observations.

Feynman Diagrams and Quantum Electrodynamics

Feynman diagrams are powerful graphical tools used in quantum field theory (QFT) to represent particle interactions and calculate scattering amplitudes. In Quantum Electrodynamics (QED), Feynman diagrams play a central role in understanding and predicting electromagnetic processes involving charged particles and photons. Let's explore how Feynman diagrams are utilized in QED.

Representation of Particle Interactions:

Feynman diagrams provide a pictorial representation of particle interactions, where particles are represented by lines and interactions by vertices. In QED, particles include electrons, positrons, and photons, while interactions involve the exchange of virtual photons.

Basic Elements of Feynman Diagrams:

1. Particles: Electrons are represented by solid lines with arrows pointing forward in time, while positrons (the antiparticles of electrons) are represented by solid lines with arrows pointing backward in time. Photons are represented by wavy lines.
2. Vertices: Vertices in Feynman diagrams represent interaction points where particles exchange momentum and energy. In QED, vertices typically involve the emission or absorption of photons by charged particles.
3. Propagators: Lines connecting vertices represent the propagation of particles between interaction points. The propagator for each particle species encodes its propagating behavior and is derived from the corresponding quantum field theory.

Rules for Constructing Feynman Diagrams:

1. Conservation Laws: Feynman diagrams must satisfy conservation laws, including conservation of charge, momentum, and energy, at each vertex.
2. Time Order: Feynman diagrams should be drawn in a way that respects the time order of particle interactions. Time flows from left to right in Feynman diagrams, with initial particles on the left and final particles on the right.
3. Loop Diagrams: Loop diagrams, where particles interact with themselves, contribute to higher-order corrections and are crucial for precise calculations in QED. Each loop in a diagram corresponds to an integration over momentum space.

Calculation of Scattering Amplitudes:

Feynman diagrams provide a systematic approach to calculating scattering amplitudes, which represent the probabilities for particle interactions to occur. The amplitude for a specific process is obtained by summing over all possible Feynman diagrams contributing to that process.

Examples of Feynman Diagrams in QED:

1. Electron-Photon Scattering: An electron emits a photon, interacts with another electron, and absorbs the photon. This process contributes to the scattering of electrons by photons.
2. Electron-Positron Annihilation: An electron and a positron collide, annihilating each other and producing a photon. This process is observed in high-energy particle collisions.

Role in Precision Calculations:

Feynman diagrams allow for precise calculations of scattering amplitudes and other observables in QED. By considering diagrams at different orders of perturbation theory, physicists can systematically include higher-order corrections and obtain results that agree with experimental measurements to extraordinary precision.

Conclusion:

Feynman diagrams are invaluable tools in Quantum Electrodynamics, providing a visual representation of particle interactions and facilitating precise calculations of scattering amplitudes and other observables. By incorporating Feynman diagrams into theoretical calculations, physicists gain insight into the underlying dynamics of electromagnetic processes and can make predictions that are tested and confirmed through experiments.

Chapter 9: Quantum Chromodynamics

Quantum Chromodynamics (QCD) is the quantum field theory that describes the strong interaction between quarks and gluons, the fundamental constituents of hadrons such as protons and neutrons. In this chapter, we explore the principles of QCD, its mathematical formalism, and its implications for understanding the structure and behavior of nuclear matter.

Foundations of Quantum Chromodynamics:

1. Color Charge: QCD is based on the concept of color charge, a property carried by quarks and gluons analogous to electric charge in electromagnetism. Quarks come in three "colors" (red, green, and blue), while gluons carry a combination of color and anti-color.
2. Asymptotic Freedom: QCD exhibits a remarkable property known as asymptotic freedom, where the strong force becomes weaker at short distances or high energies. This property allows for perturbative calculations in QCD at high energies.

Mathematical Formalism:

1. Lagrangian Density: The dynamics of QCD are described by the QCD Lagrangian density, which incorporates terms for the kinetic energy of quarks and gluons, the gluon

field strength, and the interactions between quarks and gluons. QCD Lagrangian exhibits local gauge symmetry under SU(3) color transformations.
2. Quantization of Gluon Fields: Gluons, the force carriers of the strong interaction, are quantized as vector gauge bosons. The gluon field strength tensor mediates the interaction between quarks and carries the color charge between them.

Color Confinement:

One of the key features of QCD is color confinement, which states that quarks and gluons cannot exist as isolated particles but are always confined within color-neutral bound states called hadrons. This phenomenon is responsible for the observed confinement of quarks within protons, neutrons, and other hadrons.

Lattice QCD:

Due to the nonperturbative nature of QCD at low energies, lattice QCD provides a numerical approach to solving QCD equations on a discrete space-time lattice. Lattice QCD simulations allow for the study of hadron properties, such as masses and interactions, from first principles.

Gluon Self-Interactions:

Gluons can interact with each other through self-interactions, leading to phenomena such as gluon-gluon scattering and gluon fusion. These interactions play a crucial role in understanding the dynamics of the strong force and the behavior of gluon-rich environments, such as the quark-gluon plasma.

Quark-Gluon Plasma:

At extremely high temperatures and densities, such as those found in the early universe or in heavy-ion collisions at particle accelerators, QCD predicts the existence of a phase of matter known as the quark-gluon plasma (QGP), where quarks and gluons are no longer confined within hadrons but instead form a deconfined state of matter.

Experimental Verification:

Experimental studies of QCD are carried out at particle colliders, such as the Large Hadron Collider (LHC) at CERN, where high-energy collisions between protons and heavy ions probe the properties of quarks, gluons, and hadrons. Measurements of hadron production, jet formation, and other observables provide insights into the behavior of QCD at high energies and densities.

Conclusion:

Quantum Chromodynamics stands as the theory of the strong force, providing a comprehensive description of the interactions between quarks and gluons and their consequences for the structure and behavior of nuclear matter. By elucidating the dynamics of the strong force, QCD deepens our understanding of the fundamental constituents of matter and their interactions, from the smallest scales of quarks and gluons to the formation of atomic nuclei and beyond.

The Strong Force and Quantum Chromodynamics

The strong force, also known as the strong nuclear force or color force, is one of the four fundamental forces of nature, responsible for binding quarks together to form protons, neutrons, and other hadrons. Quantum Chromodynamics (QCD) is the quantum field theory that describes the strong interaction between quarks and gluons, the force carriers of the strong force. In this section, we delve into the nature of the strong force and the principles of QCD.

Nature of the Strong Force:

1. Color Charge: Unlike the electromagnetic force, which involves electric charge, the strong force operates through a different charge called color charge. Quarks carry a color charge, which comes in three types: red, green, and blue. Gluons, the force carriers of the strong force, also carry color charge.
2. Confinement: One of the most striking features of the strong force is color confinement, which states that quarks and gluons cannot exist in isolation but are always confined within color-neutral bound states called hadrons. This phenomenon explains why free quarks have never been observed in isolation.

Principles of Quantum Chromodynamics:

1. Gauge Symmetry: QCD is based on local gauge symmetry under SU(3) color transformations. Gluons are the gauge bosons associated with this symmetry and mediate the interaction between quarks. QCD Lagrangian exhibits local gauge invariance, ensuring the consistency of the theory.
2. Asymptotic Freedom: QCD exhibits a remarkable property known as asymptotic freedom, where the strong force becomes weaker at short distances or high energies. This property allows for the perturbative calculation of QCD processes at high energies, analogous to the behavior of electromagnetism.

Mathematical Formalism:

1. Lagrangian Density: The dynamics of QCD are described by the QCD Lagrangian density, which includes terms for the kinetic energy of quarks and gluons, the gluon field strength tensor, and the interactions between quarks and gluons. QCD Lagrangian exhibits local gauge symmetry and is invariant under SU(3) color transformations.
2. Quantization of Gluon Fields: Gluons are quantized as vector gauge bosons, and the gluon field strength tensor mediates the interaction between quarks. The non-Abelian

nature of SU(3) gauge symmetry leads to the self-interactions of gluons, giving rise to rich and complex phenomena in QCD.

Experimental Verification:

Experimental studies of the strong force are carried out at particle colliders, such as the Large Hadron Collider (LHC) at CERN, where high-energy collisions between protons and heavy ions probe the properties of quarks, gluons, and hadrons. Measurements of hadron production, jet formation, and other observables provide insights into the behavior of QCD at high energies and densities.

Applications and Beyond:

QCD has applications in various areas of physics, including nuclear physics, astrophysics, and particle physics. Understanding the behavior of QCD is crucial for describing the structure of atomic nuclei, the formation of neutron stars, and the dynamics of high-energy particle collisions.

Conclusion:

The strong force and Quantum Chromodynamics play a fundamental role in our understanding of the structure and behavior of matter at the smallest scales. By elucidating the interactions between quarks and gluons and their consequences for the formation of hadrons and nuclear matter, QCD deepens our understanding of the fundamental forces of nature and their role in shaping the universe.

Quarks and Gluons

Quarks and gluons are the fundamental constituents of matter and the building blocks of protons, neutrons, and other hadrons. In the framework of Quantum Chromodynamics (QCD), quarks carry color charge and interact via the exchange of gluons, the force carriers of the strong force. Let's explore the properties and roles of quarks and gluons in more detail.

Quarks:

1. Fundamental Particles: Quarks are elementary particles that possess fractional electric charges and are the fundamental constituents of matter. There are six known types, or "flavors," of quarks: up (u), down (d), charm (c), strange (s), top (t), and bottom (b).
2. Color Charge: Quarks carry a property called color charge, which comes in three types: red, green, and blue. Each quark has a color and an associated anti-color, leading to a net color neutrality in hadrons such as protons and neutrons.

3. Confinement: Quarks are subject to the phenomenon of color confinement, which prevents isolated quarks from existing in free space. Instead, quarks are always bound together within color-neutral hadrons due to the strong force mediated by gluons.
4. Fractional Charges: Quarks possess fractional electric charges, with values of either +2/3 (up, charm, top) or -1/3 (down, strange, bottom) in units of the elementary charge. This fractional charge is a unique feature of quarks compared to other particles.

Gluons:

1. Force Carriers: Gluons are the gauge bosons that mediate the strong interaction between quarks, analogous to photons in electromagnetism. Gluons carry color charge themselves and can interact with other gluons, giving rise to the self-interactions of the strong force.
2. Color Charge: Gluons carry color charge and can exist in eight different color-anticolor combinations, corresponding to the eight generators of the SU(3) color gauge group. Gluons interact with quarks via the exchange of color charge, leading to the binding of quarks within hadrons.
3. Confinement and Self-Interaction: Gluons play a central role in color confinement, as their self-interactions give rise to the strong force field lines that bind quarks together within hadrons. The self-interactions of gluons also lead to the non-Abelian nature of QCD and the complexity of strong force dynamics.

Roles in Hadron Structure:

1. Constituent Quarks: Quarks are the constituent particles of hadrons, such as protons and neutrons. The combination of three quarks (baryons) or a quark-antiquark pair (mesons) forms color-neutral hadronic states.
2. Glueballs and Hybrid Hadrons: In addition to quarks and antiquarks, QCD allows for the formation of exotic hadronic states called glueballs, composed entirely of gluons. Hybrid hadrons, which contain both quark-antiquark pairs and gluons, are also predicted by QCD.

Experimental Studies:

Experimental studies of quarks and gluons are carried out at particle accelerators, such as the Large Hadron Collider (LHC), where high-energy collisions between protons and heavy ions probe the properties of quarks, gluons, and their interactions. These experiments provide insights into the behavior of QCD at high energies and densities, shedding light on the fundamental structure of matter.

Asymptotic Freedom and Confinement

Asymptotic freedom and confinement are two fundamental properties of the strong force, described by Quantum Chromodynamics (QCD), that govern the behavior of quarks and gluons at different energy scales. Let's explore these concepts in more detail:

Asymptotic Freedom:

1. Definition: Asymptotic freedom refers to the behavior of the strong force at short distances or high energies, where quarks and gluons interact weakly. In other words, the strength of the strong force decreases as the distance between quarks or gluons decreases, or as the energy of the interaction increases.
2. Perturbative Behavior: At high energies, QCD interactions become weak enough to be described perturbatively, similar to electromagnetism. This allows for the calculation of scattering amplitudes and other observables using perturbation theory, where higher-order corrections become increasingly important.
3. Running Coupling Constant: The strength of the strong force is characterized by a parameter called the coupling constant which depends on the energy scale of the interaction. At high energies, the coupling constant decreases logarithmically with increasing energy, reflecting the asymptotic freedom of QCD.
4. Experimental Evidence: Asymptotic freedom has been confirmed through experimental studies, such as deep inelastic scattering experiments and high-energy collisions at particle accelerators like the Large Hadron Collider (LHC). These experiments have provided evidence for the weak coupling of quarks and gluons at short distances.

Confinement:

1. Definition: Confinement is the phenomenon whereby quarks and gluons are permanently confined within color-neutral bound states called hadrons, such as protons, neutrons, and mesons. Despite the weak interaction at short distances, quarks and gluons cannot exist in isolation but are always confined within hadrons.
2. Color Neutrality: Confinement arises from the strong force between quarks, which becomes stronger as quarks move farther apart. The force field lines between quarks form tubes or flux tubes, which store energy and prevent the separation of quarks. As a result, only color-neutral combinations of quarks can exist as stable particles.
3. Lattice QCD and Glueballs: The phenomenon of confinement is a nonperturbative aspect of QCD that is challenging to study analytically. Numerical simulations using lattice QCD techniques provide insight into the nature of confinement and the formation of hadronic bound states. These simulations predict the existence of exotic hadrons called glueballs, composed entirely of gluons.
4. Experimental Challenges: While confinement has not been directly observed in experiments due to the inability to isolate quarks and gluons, indirect evidence for confinement comes from the absence of free quarks in nature and the spectrum of hadronic bound states observed in high-energy collisions.

Dual Description:

1. Dual Descriptions: Confinement can be understood through a dual description in terms of string theory, where the strong force is described by the dynamics of string-like objects known as flux tubes. In this picture, the confinement of quarks is analogous to the confinement of the endpoints of a string.
2. Quark-Gluon Plasma: At extreme temperatures and densities, such as those achieved in heavy-ion collisions at particle accelerators, QCD predicts the existence of a deconfined phase of matter called the quark-gluon plasma (QGP), where quarks and gluons are no longer confined within hadrons but instead form a strongly interacting plasma.

Conclusion:

Asymptotic freedom and confinement are two fundamental aspects of the strong force described by Quantum Chromodynamics. While asymptotic freedom governs the weak coupling of quarks and gluons at short distances or high energies, confinement explains why quarks and gluons are permanently confined within color-neutral hadronic bound states. Together, these properties provide a comprehensive understanding of the behavior of quarks and gluons and their role in the structure of matter.

Chapter 10: Quantum Gravity and Unified Theories

Quantum gravity is a theoretical framework that aims to reconcile the principles of general relativity with those of quantum mechanics, providing a consistent description of gravity at the smallest scales of space and time. Unified theories, also known as Grand Unified Theories (GUTs) or theories of everything (TOEs), seek to unify all fundamental forces of nature into a single theoretical framework. In this chapter, we explore the challenges and prospects of quantum gravity and unified theories.

Challenges of Quantum Gravity:

1. Incompatibility of Quantum Mechanics and General Relativity: General relativity describes gravity as the curvature of spacetime, while quantum mechanics describes the behavior of particles and fields at the quantum level. Combining these two theories leads to conceptual and mathematical challenges, particularly in regions of extreme curvature or high energy density.
2. Quantization of Gravity: The quantization of gravity involves treating the gravitational field as a quantum field, similar to other fundamental forces. However, the nonrenormalizability of gravity poses challenges for straightforward quantization methods, requiring the development of new mathematical techniques and frameworks.

3. Nature of Spacetime: Quantum gravity theories must address the fundamental nature of spacetime at the Planck scale, where quantum effects become significant. This includes understanding the structure of spacetime at extremely small distances and the emergence of spacetime from more fundamental constituents.

Approaches to Quantum Gravity:

1. String Theory: String theory is a theoretical framework that models elementary particles as one-dimensional strings instead of point particles. In string theory, gravity emerges naturally as the dynamics of strings propagating in higher-dimensional spacetimes. String theory offers a promising candidate for a consistent theory of quantum gravity.
2. Loop Quantum Gravity: Loop quantum gravity is an approach to quantum gravity that quantizes the geometry of spacetime directly. In this framework, spacetime is discretized into elementary units, and physical quantities are represented by operators acting on a space of states. Loop quantum gravity provides insights into the quantum nature of spacetime.
3. Emergent Gravity: Some approaches to quantum gravity propose that gravity emerges from more fundamental degrees of freedom, such as entanglement or quantum information. In this view, gravity is not fundamental but arises as an effective description of underlying quantum phenomena.

Unified Theories:

1. Grand Unified Theories (GUTs): GUTs aim to unify the electromagnetic, weak, and strong nuclear forces into a single gauge theory based on a larger symmetry group. GUTs predict the existence of new particles and interactions beyond those observed in the Standard Model of particle physics.
2. Theories of Everything (TOEs): TOEs seek to unify all fundamental forces, including gravity, into a single theoretical framework. String theory and M-theory are examples of TOEs that attempt to describe all known particles and interactions within a unified mathematical framework.

Experimental Signatures:

1. Particle Accelerators: Experimental tests of unified theories and quantum gravity are challenging due to the extremely high energies and small distances involved. Particle accelerators such as the Large Hadron Collider (LHC) may provide indirect evidence for new particles predicted by unified theories.
2. Cosmological Observations: Observations of the cosmic microwave background, gravitational waves, and the large-scale structure of the universe offer insights into the behavior of gravity and the early universe, providing constraints on quantum gravity theories and unified models.

Conclusion:

Quantum gravity and unified theories represent ambitious endeavors to unify our understanding of the fundamental forces of nature at the most fundamental level. While progress has been made in theoretical developments and experimental observations, significant challenges remain in reconciling the principles of quantum mechanics and general relativity and in testing predictions of unified theories at the energies accessible to current experiments. Nonetheless, these efforts hold the promise of uncovering profound insights into the nature of spacetime, matter, and the universe at its most fundamental level.

Challenges of Quantum Gravity

Quantum gravity is a theoretical framework that aims to reconcile the principles of quantum mechanics with those of general relativity, providing a consistent description of gravity at the smallest scales of space and time. However, achieving this goal presents several formidable challenges that remain open areas of research. Let's explore some of the key challenges faced by quantum gravity theorists:

1. Nonrenormalizability:

One of the major challenges in quantizing gravity is its nonrenormalizability. Unlike other fundamental forces, such as electromagnetism and the strong nuclear force, the gravitational interaction becomes infinitely strong at small distances or high energies. This poses difficulties in the standard methods of renormalization used in quantum field theory, leading to infinities that cannot be removed through renormalization procedures.

2. Planck Scale Physics:

Quantum gravity becomes significant at the Planck scale, where the effects of quantum mechanics and gravity become comparable. At this scale, spacetime is expected to be highly curved and fluctuating, leading to a breakdown of classical spacetime geometry. Understanding the quantum behavior of spacetime and the emergence of classical spacetime from more fundamental degrees of freedom is a key challenge in quantum gravity.

3. Unification of Forces:

Quantum gravity seeks to unify gravity with the other fundamental forces of nature—electromagnetism, the weak nuclear force, and the strong nuclear force—into a single theoretical framework. Achieving such unification requires a deep understanding of the underlying symmetries and structures that govern the behavior of these forces at high energies. Developing a unified theory that encompasses all known interactions remains a major theoretical challenge.

4. Black Hole Information Paradox:

The study of black holes poses fundamental challenges to our understanding of quantum gravity. The existence of black hole singularities and the loss of information during black hole evaporation, as predicted by Hawking radiation, raise questions about the compatibility of quantum mechanics with the classical description of black holes. Resolving the black hole information paradox requires a quantum theory of gravity that can describe the behavior of matter and information in the presence of strong gravitational fields.

5. Experimental Verification:

Experimental verification of quantum gravity theories is challenging due to the extreme energies and distances involved. Current experimental techniques are not yet capable of probing the Planck scale, where quantum gravity effects are expected to become significant. Developing new experimental techniques and observations that can test predictions of quantum gravity theories remains an important goal for the field.

6. Consistency with Observations:

Any successful theory of quantum gravity must be consistent with observational data from astrophysical observations, particle physics experiments, and cosmological observations. Ensuring that quantum gravity predictions match observational constraints and accurately describe phenomena observed in nature is essential for validating the theory.

Conclusion:

Quantum gravity represents one of the most significant challenges in theoretical physics, aiming to unify our understanding of gravity with quantum mechanics. While progress has been made in various approaches to quantum gravity, such as string theory, loop quantum gravity, and emergent gravity, many fundamental questions remain unanswered. Addressing these challenges requires interdisciplinary collaboration between theoretical physicists, mathematicians, and experimentalists to push the boundaries of our understanding of the universe at its most fundamental level.

String Theory

String theory is a theoretical framework in theoretical physics that attempts to describe all fundamental particles and forces of nature in terms of one-dimensional objects called strings. It is a leading candidate for a theory of quantum gravity, aiming to reconcile the principles of general relativity and quantum mechanics. Let's delve deeper into the key aspects of string theory:

1. Fundamental Idea:

In string theory, the fundamental building blocks of the universe are not point-like particles but rather one-dimensional strings, which can vibrate in different modes. The vibrational patterns of these strings determine the properties and behavior of particles, such as mass, charge, and spin. The theory posits that all particles observed in nature, including quarks, leptons, and gauge bosons, arise from different vibrational modes of strings.

2. String Dynamics:

The dynamics of strings are governed by the Nambu-Goto action or the Polyakov action, depending on the formulation of the theory. These actions describe how strings move through spacetime and interact with each other. By quantizing the dynamics of strings, one can derive the equations of motion and calculate scattering amplitudes for various particle interactions.

3. Extra Dimensions:

String theory requires the existence of extra spatial dimensions beyond the familiar three dimensions of space and one dimension of time. These extra dimensions are compactified or curled up at very small scales, making them invisible to our everyday perception. The number and shape of these extra dimensions play a crucial role in determining the properties of particles and forces in string theory.

4. Supersymmetry:

Supersymmetry is a key feature of many formulations of string theory. It postulates a symmetry between fermions (particles with half-integer spin) and bosons (particles with integer spin). Supersymmetry predicts the existence of superpartners for each known particle, providing a natural candidate for dark matter and addressing certain theoretical issues in particle physics, such as the hierarchy problem.

5. Different String Theories:

String theory encompasses several different formulations, including Type I, Type IIA, Type IIB, heterotic SO(32), and heterotic E8×E8. These formulations differ in their choice of string type (open or closed) and their treatment of supersymmetry and extra dimensions. The five consistent superstring theories were later shown to be different limits of a more fundamental theory called M-theory.

6. M-theory:

M-theory is a conjectured theory that underlies and unifies the various string theories. It is believed to be a more fundamental framework that encompasses all known string theories as different limits or approximations. M-theory is still under active research, and its precise formulation and implications remain topics of ongoing investigation.

7. Challenges and Controversies:

Despite its promise, string theory faces several challenges and controversies. These include the difficulty in obtaining testable predictions, the existence of a vast landscape of possible vacuum states, the lack of experimental verification, and the problem of nonuniqueness in selecting a vacuum state. Additionally, the high energy scales required for probing string theory effects remain beyond the reach of current experimental capabilities.

Conclusion:

String theory represents a profound and ambitious attempt to unify the fundamental forces of nature within a single theoretical framework. While it offers tantalizing possibilities for resolving long-standing problems in theoretical physics, such as the quantization of gravity and the unification of forces, it also presents significant theoretical and conceptual challenges. Continued research into string theory and related areas of theoretical physics promises to shed light on the fundamental nature of the universe and our place within it.

Loop Quantum Gravity and Other Approaches

Loop Quantum Gravity (LQG) is a theoretical framework in theoretical physics that aims to quantize gravity, providing a consistent quantum description of spacetime at the smallest scales. Unlike string theory, which posits the existence of fundamental one-dimensional strings, LQG treats spacetime itself as quantized, breaking it down into discrete units or "loops." Let's explore Loop Quantum Gravity and compare it with other approaches to quantum gravity:

Loop Quantum Gravity (LQG):

1. Quantization of Geometry: LQG approaches the quantization of gravity by quantizing the geometry of spacetime directly. In this framework, spacetime is discretized into elementary units known as "quantum states" or "spin networks," which represent the possible configurations of geometry.
2. Loop Variables: The fundamental variables of LQG are loops or one-dimensional curves embedded in the three-dimensional spatial manifold. These loops represent the spatial geometry and are quantized using techniques from mathematical quantum theory, such as noncommutative geometry.
3. Area and Volume Quantization: In LQG, geometric quantities such as area and volume have discrete spectra, meaning they can only take on certain quantized values. This discrete structure of spacetime is believed to resolve the problem of ultraviolet divergences encountered in the quantization of gravity.

4. Black Hole Entropy and Cosmology: LQG has made significant contributions to our understanding of black hole entropy and cosmology. The quantization of black hole horizons leads to a discrete spectrum of area, consistent with the Bekenstein-Hawking entropy formula. LQG also offers insights into the early universe and the resolution of cosmological singularities.

Other Approaches to Quantum Gravity:

1. String Theory: String theory is another leading candidate for a theory of quantum gravity. Unlike LQG, which treats spacetime geometry as discrete, string theory postulates the existence of fundamental one-dimensional strings as the building blocks of the universe. String theory aims to reconcile gravity with quantum mechanics by quantizing the behavior of strings propagating in higher-dimensional spacetimes.
2. Asymptotic Safety: Asymptotic safety is a proposal for quantizing gravity within the framework of quantum field theory. It suggests that gravity may be renormalizable at a nonperturbative level, meaning that quantum gravitational effects can be consistently described without encountering infinite quantities. This approach relies on the existence of a non-Gaussian fixed point in the renormalization group flow of gravity.
3. Causal Dynamical Triangulations (CDT): CDT is a lattice-based approach to quantum gravity that discretizes spacetime into simplices and evolves them dynamically according to certain causal constraints. By summing over all possible triangulations of spacetime, CDT aims to obtain a path integral formulation of quantum gravity. CDT has been used to study the emergence of classical spacetime from quantum fluctuations and the behavior of spacetime near the Planck scale.

Comparison:

1. Quantization Approach: LQG quantizes the geometry of spacetime directly, while string theory quantizes the behavior of fundamental strings within a continuous spacetime background.
2. Discreteness vs. Continuity: LQG predicts a discrete structure of spacetime at the Planck scale, while string theory posits a continuous spacetime background with strings propagating within it.
3. Experimental Predictions: Both LQG and string theory face challenges in making testable predictions that can be verified experimentally due to the high energy scales involved. However, they offer valuable insights into the nature of spacetime and the unification of fundamental forces.

Conclusion:

Loop Quantum Gravity, along with other approaches such as string theory, asymptotic safety, and causal dynamical triangulations, represents a diverse set of theoretical frameworks aimed at understanding the quantum nature of gravity. While each approach has its own strengths and challenges, they collectively contribute to our quest for a unified theory of quantum gravity and a deeper understanding of the fundamental structure of the universe. Ongoing research in these areas promises to illuminate the nature of spacetime and the fundamental forces at play in the cosmos.

Chapter 11: Quantum Cosmology

Quantum cosmology is a branch of theoretical physics that applies the principles of quantum mechanics to the study of the universe as a whole. It seeks to address fundamental questions about the origin, evolution, and fate of the cosmos by treating the universe as a quantum system. In this chapter, we explore the key concepts and developments in quantum cosmology.

1. Introduction to Quantum Cosmology:

1.1 Foundations: Quantum cosmology seeks to provide a quantum description of the universe, incorporating principles from quantum mechanics and general relativity. It addresses questions such as the initial conditions of the universe, the nature of the Big Bang singularity, and the quantum behavior of spacetime.
1.2 Wave Function of the Universe: A central concept in quantum cosmology is the wave function of the universe, which describes the quantum state of the entire cosmos. The wave function encodes information about the possible configurations of the universe's geometry and matter content.

2. Early Universe Models:

2.1 Inflationary Cosmology: Inflationary models propose that the universe underwent a period of rapid expansion in its early history, driven by a scalar field called the inflaton. Quantum fluctuations during inflation are thought to give rise to the seeds of cosmic structure observed in the universe today.
2.2 Quantum Cosmological Models: Quantum cosmology explores the behavior of the universe at very early times, near the Big Bang singularity. Models such as the Wheeler-DeWitt equation and loop quantum cosmology offer approaches to quantizing the gravitational field and studying the quantum evolution of the universe.

3. Quantum Gravity Effects:

3.1 Planck Scale Physics: Quantum cosmology probes the behavior of the universe at the Planck scale, where the effects of quantum gravity become significant. At these energies, the discrete nature of spacetime and the quantization of geometry are expected to play a crucial role.
3.2 Primordial Gravitational Waves: Quantum cosmological models predict the existence of primordial gravitational waves, which are ripples in the fabric of spacetime generated during the early moments of the universe. These gravitational waves carry valuable information about the

conditions of the early universe and may be detected through experiments such as cosmic microwave background polarization measurements.

4. Observational Implications:

4.1 Cosmic Microwave Background: Observations of the cosmic microwave background (CMB) radiation provide important constraints on quantum cosmological models. Anisotropies and polarization patterns in the CMB offer insights into the initial conditions and evolution of the universe.
4.2 Large-Scale Structure: The distribution of galaxies and cosmic structures in the universe reflects the imprint of quantum fluctuations during the early universe. Studying the large-scale structure of the cosmos can test predictions of quantum cosmological models and provide clues about the underlying quantum nature of the universe.

5. Future Directions:

5.1 Experimental Tests: Advancements in observational cosmology and high-energy physics may offer opportunities to test predictions of quantum cosmological models. Future experiments, such as gravitational wave detectors and precision measurements of the CMB, hold promise for probing the quantum behavior of the universe.
5.2 Unified Theories: Quantum cosmology aims to contribute to the quest for a unified theory of quantum gravity, which reconciles quantum mechanics with general relativity. By studying the quantum behavior of the universe, researchers seek to uncover deeper insights into the fundamental nature of spacetime and the laws governing the cosmos.

Conclusion:

Quantum cosmology represents a frontier in theoretical physics, where the principles of quantum mechanics and general relativity converge to explore the origins and evolution of the universe. By applying quantum concepts to the study of the cosmos, quantum cosmology offers a framework for addressing some of the deepest questions about the nature of existence and the fundamental structure of reality. Continued research in this field promises to deepen our understanding of the universe's past, present, and future, and to unlock new insights into the nature of quantum gravity and the origin of cosmic structure.

Quantum Origins of the Universe

The study of the quantum origins of the universe delves into the fundamental questions of how the cosmos began and what physical processes governed its earliest moments. Quantum cosmology, a branch of theoretical physics, aims to provide insights into these questions by applying the principles of quantum mechanics to the study of the universe as a whole. Let's explore the quantum origins of the universe and the key concepts involved:

1. Pre-Big Bang Cosmology:

1.1 Primordial Quantum Fluctuations: Quantum fluctuations in the early universe are believed to have played a crucial role in seeding the formation of cosmic structure. These fluctuations, originating from quantum uncertainty, left an imprint on the cosmic microwave background radiation observed today.

1.2 Quantum Gravity Regime: In the Planck epoch, when the universe was extremely hot and dense, quantum gravity effects dominated. At these scales, the discrete nature of spacetime and the quantization of geometry become significant, necessitating a quantum description of the universe.

2. Inflationary Cosmology:

2.1 Inflationary Epoch: Inflationary models propose that the universe underwent a period of exponential expansion in its early moments, driven by a scalar field called the inflaton. Quantum fluctuations during inflation stretched across cosmic scales, providing the seeds for the formation of galaxies and large-scale structure.

2.2 Quantum Fluctuations and Density Perturbations: Quantum fluctuations in the inflaton field led to variations in the energy density of the early universe. These density perturbations, imprinted as temperature fluctuations in the cosmic microwave background, are observed as the tiny temperature anisotropies in the CMB.

3. Quantum Cosmological Models:

3.1 Wheeler-DeWitt Equation: The Wheeler-DeWitt equation is a key equation in quantum cosmology that describes the wave function of the universe. It represents the quantum analog of the classical Hamiltonian constraint in general relativity and encapsulates the quantum dynamics of the universe.

3.2 Loop Quantum Cosmology: Loop quantum cosmology extends loop quantum gravity techniques to cosmological scenarios. It provides a framework for quantizing the gravitational field and studying the quantum evolution of the universe, particularly near the Big Bang singularity.

4. Cosmic Microwave Background (CMB):

4.1 Quantum Fluctuations in the CMB: Observations of the cosmic microwave background radiation reveal temperature fluctuations at the level of one part in a hundred thousand. These fluctuations are thought to arise from quantum fluctuations in the early universe, providing valuable insights into its quantum origins.

4.2 Baryon Acoustic Oscillations: Baryon acoustic oscillations (BAOs) in the large-scale distribution of galaxies also reflect the quantum seeds of cosmic structure. These oscillations result from quantum pressure waves in the early universe, imprinted in the distribution of matter and visible in galaxy surveys.

5. Future Directions:

5.1 Experimental Tests: Advancements in observational cosmology, gravitational wave astronomy, and particle physics may offer opportunities to test predictions of quantum cosmological models. Future experiments, such as high-precision measurements of the CMB and gravitational wave detectors, hold promise for probing the quantum nature of the universe.
5.2 Unified Theories: Quantum cosmology aims to contribute to the quest for a unified theory of quantum gravity, which reconciles quantum mechanics with general relativity. By studying the quantum behavior of the universe's earliest moments, researchers seek to uncover deeper insights into the fundamental nature of spacetime and the laws governing the cosmos.

Conclusion:

The study of the quantum origins of the universe represents a frontier in theoretical physics, where the principles of quantum mechanics and general relativity converge to explore the earliest moments of cosmic history. By applying quantum concepts to the study of the cosmos, quantum cosmology offers a framework for addressing some of the deepest questions about the nature of existence and the fundamental structure of reality. Continued research in this field promises to deepen our understanding of the universe's origins and evolution, revealing new insights into the quantum nature of spacetime and the origin of cosmic structure.

Inflationary Cosmology

Inflationary cosmology is a theoretical framework in cosmology that proposes the universe underwent a rapid exponential expansion in its earliest moments, known as cosmic inflation. This period of accelerated expansion is thought to have occurred shortly after the Big Bang and provides a solution to several outstanding problems in cosmology. Let's delve into the key concepts and implications of inflationary cosmology:

1. The Inflationary Paradigm:

1.1 Motivation: Inflationary cosmology was proposed to address several puzzles in the standard Big Bang model, such as the horizon problem, flatness problem, and origin of structure. By positing a period of rapid expansion, inflationary models provide mechanisms for resolving these issues and explaining the large-scale properties of the universe.
1.2 Exponential Expansion: During inflation, the scale factor of the universe expanded exponentially, causing space to stretch rapidly. This expansion occurs faster than the speed of light, allowing distant regions of the universe to come into causal contact and reach thermal equilibrium, thus resolving the horizon problem.

2. Inflationary Models:

2.1 Inflaton Field: Inflation is typically driven by the dynamics of a scalar field called the inflaton. The inflaton field is characterized by a potential energy density that drives the exponential expansion of the universe. The dynamics of the inflaton field determine the duration and properties of inflation.

2.2 Quantum Fluctuations: Quantum fluctuations in the inflaton field during inflation are amplified by the expansion of space and lead to variations in the energy density of the early universe. These fluctuations serve as the seeds for the formation of cosmic structure, including galaxies, clusters, and large-scale structure.

3. Implications and Predictions:

3.1 Flatness and Homogeneity: Inflationary models predict that the universe should be spatially flat on large scales, consistent with observations from cosmic microwave background (CMB) radiation and galaxy surveys. The exponential expansion during inflation smooths out curvature and anisotropies, leading to a homogeneous and isotropic universe.

3.2 Density Perturbations: Quantum fluctuations during inflation generate density perturbations, which are observed as temperature anisotropies in the cosmic microwave background. These density perturbations are nearly scale-invariant and Gaussian, providing a mechanism for the observed structure in the universe.

4. Observational Tests:

4.1 Cosmic Microwave Background (CMB): Measurements of the cosmic microwave background radiation provide strong support for inflationary cosmology. Observations of temperature fluctuations in the CMB are consistent with the predictions of inflationary models, including the scale-invariant spectrum of density perturbations.

4.2 Baryon Acoustic Oscillations (BAOs): Large-scale structure observations, such as galaxy surveys, also support the predictions of inflationary cosmology. Baryon acoustic oscillations in the distribution of galaxies reflect the imprint of density perturbations generated during inflation, confirming the role of quantum fluctuations in structure formation.

5. Open Questions and Future Directions:

5.1 Quantum Gravity and Initial Conditions: The origin of the inflaton field and the precise mechanism of inflation remain open questions. Understanding the quantum nature of gravity and the initial conditions for inflationary models are areas of active research in theoretical cosmology.

5.2 Extensions and Alternatives: While inflationary cosmology has been successful in addressing many observational challenges, alternative models and extensions to inflation are also being explored. These include multifield inflation, modified gravity theories, and bouncing cosmologies, which offer different mechanisms for generating primordial perturbations.

Conclusion:

Inflationary cosmology has emerged as a leading paradigm for understanding the early universe and the origin of cosmic structure. By positing a period of rapid exponential expansion, inflationary models provide elegant solutions to longstanding puzzles in cosmology and make testable predictions that are consistent with observations. Continued advancements in observational cosmology, theoretical physics, and experimental techniques promise to further refine our understanding of inflation and its implications for the nature of the universe.

Quantum Cosmological Models

Quantum cosmological models are theoretical frameworks that apply the principles of quantum mechanics to the study of the entire universe. These models aim to describe the quantum behavior of the universe, including its origin, evolution, and ultimate fate. Here are some key quantum cosmological models and their features:

1. Wheeler-DeWitt Equation:

1.1 Foundations: The Wheeler-DeWitt equation is a central equation in quantum cosmology, proposed by Bryce DeWitt and John Archibald Wheeler in the 1960s. It is derived from the canonical quantization of general relativity and describes the wave function of the universe.
1.2 Wave Function of the Universe: The Wheeler-DeWitt equation represents the quantum state of the entire universe. It is a functional differential equation that encapsulates the dynamics of spacetime geometry and matter fields, without reference to an external time parameter.
1.3 Quantum Constraints: The Wheeler-DeWitt equation arises from the Hamiltonian constraint of general relativity, which states that the total energy of the universe is zero. This constraint leads to a wave function that satisfies the quantum version of Einstein's field equations.

2. Loop Quantum Cosmology (LQC):

2.1 Quantization of Geometry: Loop quantum cosmology extends techniques from loop quantum gravity to cosmological scenarios. It quantizes the gravitational field and the geometry of spacetime, leading to a discrete structure of spacetime at the Planck scale.
2.2 Singularity Resolution: One of the key features of LQC is its ability to resolve the classical Big Bang singularity. Quantum effects prevent the universe from collapsing to a point of infinite density, leading to a bounce and the emergence of a new expanding phase.
2.3 Quantum Bounce: In LQC, the universe undergoes a quantum bounce instead of a classical singularity. This bounce occurs when the energy density of the universe reaches the Planck scale, leading to repulsive gravitational effects that prevent gravitational collapse.

3. Euclidean Quantum Gravity:

3.1 Path Integral Formulation: Euclidean quantum gravity approaches the quantum description of the universe using the path integral formulation. It sums over all possible spacetime geometries, weighted by the exponential of the Einstein-Hilbert action.

3.2 Imaginary Time: In Euclidean quantum gravity, spacetime is analytically continued to imaginary time, which allows for a Wick rotation from Lorentzian spacetime to Euclidean spacetime. This transformation simplifies the mathematical description of quantum gravity.

4. Hartle-Hawking State:

4.1 No-Boundary Proposal: The Hartle-Hawking state is a proposal for the wave function of the universe based on the no-boundary condition. According to this proposal, the universe has no singular boundary in imaginary time, and the wave function is well-defined for all spacetime geometries.

4.2 Wave Function Amplitude: The Hartle-Hawking wave function assigns amplitudes to different possible spacetime geometries, weighted by the Euclidean action. It describes a superposition of closed, compact, and boundaryless universes, with no preferred initial or final state.

5. Observational Implications:

5.1 Cosmic Microwave Background (CMB): Quantum cosmological models make predictions about the statistical properties of the cosmic microwave background radiation, such as the angular power spectrum and polarization patterns. Observations of the CMB provide valuable constraints on these models.

5.2 Large-Scale Structure: The distribution of galaxies and cosmic structures in the universe also reflects the quantum nature of the initial conditions and the evolution of cosmological perturbations. Studying the large-scale structure of the cosmos can test predictions of quantum cosmological models and provide insights into the early universe.

Conclusion:

Quantum cosmological models offer theoretical frameworks for describing the quantum behavior of the entire universe, from its earliest moments to its present state. These models provide insights into the origin, evolution, and fate of the cosmos, addressing fundamental questions in cosmology and theoretical physics. By combining principles from quantum mechanics and general relativity, quantum cosmology aims to uncover the underlying laws governing the universe and deepen our understanding of its fundamental nature.

Chapter 12: Quantum Philosophy and Interpretations

Quantum physics revolutionized our understanding of the fundamental nature of reality, challenging classical intuitions and posing profound philosophical questions about the nature of existence and the role of observers in shaping reality. In this chapter, we explore the philosophical implications and various interpretations of quantum mechanics:

1. Copenhagen Interpretation:

1.1 Foundations: The Copenhagen interpretation, proposed by Niels Bohr and Werner Heisenberg in the 1920s, is one of the earliest interpretations of quantum mechanics. It emphasizes the role of the observer and the idea of wavefunction collapse upon measurement.
1.2 Wave-Particle Duality: According to the Copenhagen interpretation, particles such as electrons exist in a state of superposition, simultaneously exhibiting wave-like and particle-like properties. Only upon measurement do they collapse into definite states.

2. Many-Worlds Interpretation:

2.1 Parallel Universes: The Many-Worlds interpretation, proposed by Hugh Everett III in the 1950s, suggests that every quantum event results in the branching of the universe into multiple parallel realities, each corresponding to a different outcome of the measurement.
2.2 Quantum Superposition: In the Many-Worlds interpretation, quantum superposition is taken to its logical conclusion, with all possible outcomes of a measurement coexisting in separate branches of the universal wavefunction. Observers themselves become entangled with the quantum system, experiencing a subjective reality within their branch.

3. Pilot Wave Theory (De Broglie-Bohm Theory):

3.1 Deterministic Realism: Pilot wave theory, also known as de Broglie-Bohm theory, proposes a deterministic interpretation of quantum mechanics. It posits the existence of a guiding wave that determines the trajectory of particles, while retaining the wave-particle duality of quantum mechanics.
3.2 Hidden Variables: In pilot wave theory, the apparent randomness of quantum measurements arises from our ignorance of the exact initial conditions and the hidden variables that determine the behavior of particles. Unlike in the Copenhagen interpretation, particles have well-defined trajectories at all times.

4. Quantum Bayesianism (QBism):

4.1 Subjective Interpretation: Quantum Bayesianism, or QBism, emphasizes the subjective nature of quantum probabilities and the role of observers in constructing their own personal probabilities based on their experiences and interactions with the quantum world.

4.2 Bayesian Probability: In QBism, quantum probabilities are interpreted as degrees of belief or subjective judgments about the outcomes of measurements. Observers update their probabilities in a Bayesian manner, incorporating new evidence and refining their understanding of reality.

5. Interpretational Challenges and Open Questions:

5.1 Measurement Problem: The interpretation of quantum mechanics raises fundamental questions about the nature of measurement and the role of observers in collapsing the wavefunction. Different interpretations offer contrasting views on the nature of reality and the resolution of the measurement problem.
5.2 Quantum Entanglement: The phenomenon of quantum entanglement, where particles become correlated in such a way that the state of one particle instantaneously influences the state of another, poses challenges to our understanding of locality and causality. Various interpretations offer different explanations for the nature of entanglement.

6. Philosophical Implications:

6.1 Reality and Observation: Quantum mechanics blurs the distinction between the observer and the observed, challenging traditional notions of objectivity and realism. The philosophical implications of quantum mechanics extend to questions about the nature of reality, consciousness, and free will.
6.2 Epistemological Limits: Quantum mechanics highlights the epistemological limits of scientific knowledge and the inherent uncertainty and indeterminacy of the quantum world. Philosophical reflections on these limits offer insights into the nature of scientific inquiry and the boundaries of human understanding.

Conclusion:

Quantum philosophy and interpretations of quantum mechanics offer diverse perspectives on the nature of reality, the role of observers, and the fundamental principles governing the quantum world. By exploring these interpretations and their philosophical implications, we deepen our understanding of the profound mysteries of quantum mechanics and our place within the quantum cosmos. Continued philosophical inquiry and interdisciplinary dialogue are essential for grappling with the profound philosophical questions raised by quantum physics and advancing our understanding of the nature of existence.

Copenhagen Interpretation

The Copenhagen interpretation is one of the foundational interpretations of quantum mechanics, proposed by Niels Bohr and Werner Heisenberg in the 1920s. It provides a framework for understanding the mathematical formalism of quantum mechanics and the philosophical implications of its probabilistic nature. Here's a closer look at the key principles and features of the Copenhagen interpretation:

1. Wave-Particle Duality:

1.1 Complementarity: One of the central tenets of the Copenhagen interpretation is the principle of complementarity, which asserts that particles can exhibit both wave-like and particle-like behavior depending on the experimental context. For example, in the double-slit experiment, particles behave like waves when unobserved but like particles when their paths are measured.
1.2 Superposition: According to the Copenhagen interpretation, particles exist in a state of superposition, meaning they can simultaneously occupy multiple states or positions until they are observed or measured. This superposition of states is described by the wave function, which evolves deterministically according to the Schrödinger equation until a measurement occurs.

2. Wavefunction Collapse:

2.1 Measurement Problem: The Copenhagen interpretation addresses the measurement problem, which arises from the apparent collapse of the wave function upon measurement. When a measurement is made, the wave function collapses to a single eigenstate corresponding to the observed outcome, with probabilities given by the Born rule.
2.2 Observer Effect: In the Copenhagen interpretation, the act of measurement is intimately tied to the presence of an observer or a measuring apparatus. Observers interact with the quantum system, causing its wave function to collapse and yielding a definite measurement outcome.

3. Epistemological Approach:

3.1 Probabilistic Interpretation: The Copenhagen interpretation adopts a probabilistic interpretation of quantum mechanics, emphasizing the role of probability amplitudes and statistical predictions in describing the behavior of quantum systems. It views quantum mechanics as providing a predictive framework for calculating the probabilities of measurement outcomes.
3.2 No Ontological Reality: Unlike some realist interpretations of quantum mechanics, the Copenhagen interpretation does not attribute ontological reality to the quantum state or wave function. Instead, it treats the wave function as a mathematical tool for predicting measurement outcomes, without implying the existence of underlying hidden variables.

4. Criticisms and Controversies:

4.1 Subjectivity: Critics of the Copenhagen interpretation have raised concerns about its subjective elements, such as the role of observers and the lack of a clear criterion for when wave function collapse occurs. Some argue that the interpretation lacks a precise ontology and relies too heavily on anthropocentric concepts.
4.2 Incomplete Description: Another criticism is that the Copenhagen interpretation provides an incomplete description of quantum reality, leaving unanswered questions about the nature of measurement and the underlying dynamics of quantum systems. Alternative interpretations,

such as the Many-Worlds interpretation and pilot wave theory, offer different perspectives on these issues.

Conclusion:

The Copenhagen interpretation of quantum mechanics remains one of the most widely taught and discussed interpretations in the field. While it has faced criticism and sparked debate among physicists and philosophers, it continues to provide a foundational framework for understanding the probabilistic nature of quantum phenomena and the role of measurement in shaping reality. By emphasizing the complementarity of wave and particle properties and the probabilistic nature of quantum systems, the Copenhagen interpretation offers valuable insights into the mysterious world of quantum mechanics.

Many-Worlds Interpretation

The Many-Worlds interpretation (MWI) is a prominent interpretation of quantum mechanics that offers a radical and counterintuitive view of reality. Proposed by Hugh Everett III in the 1950s, MWI suggests that every quantum event results in the splitting of the universe into multiple parallel realities, or "worlds," each corresponding to a different outcome of the measurement. Here's an exploration of the key principles and features of the Many-Worlds interpretation:

1. Parallel Universes:

1.1 Branching of Reality: According to the Many-Worlds interpretation, whenever a quantum measurement is made, the universe branches into multiple non-communicating branches, each representing a different possible outcome of the measurement. These branches exist in parallel, constituting an ever-expanding "multiverse" of alternate realities.
1.2 Quantum Superposition: MWI maintains that particles exist in a state of superposition, simultaneously occupying multiple states until a measurement is made. Rather than collapsing to a single outcome, the wave function evolves deterministically, with each possible outcome realized in a separate branch of the multiverse.

2. Schrödinger's Cat and Quantum Indeterminacy:

2.1 Illustrative Example: The famous thought experiment known as Schrödinger's cat is often used to illustrate the implications of MWI. In this scenario, a cat in a sealed box is both alive and dead until the box is opened, at which point the observer perceives one outcome while another outcome is realized in a parallel universe.
2.2 Resolution of Measurement Problem: MWI resolves the measurement problem by denying the need for wave function collapse. Instead, the wave function evolves unitarily, encompassing all possible outcomes of a measurement. Observers themselves become entangled with the quantum system, experiencing subjective outcomes within their respective branches.

3. Quantum Decoherence:

3.1 Emergence of Classical Reality: MWI relies on the concept of quantum decoherence to explain the apparent emergence of classical reality from the underlying quantum substrate. Decoherence occurs when quantum systems interact with their environment, leading to the suppression of interference between different branches and the emergence of classical-like behavior.

3.2 Consistency and Conservation Laws: Despite the proliferation of parallel branches, MWI maintains the conservation of energy, momentum, and other physical quantities across all branches. Each branch evolves independently according to deterministic laws, ensuring consistency and coherence within the multiverse.

4. Observational Implications:

4.1 Quantum Interference Patterns: MWI predicts that interference patterns observed in double-slit and other quantum experiments result from the coherent superposition of different branches of the multiverse. Each branch contributes to the overall interference pattern, giving rise to the observed probabilistic outcomes.

4.2 Experimental Tests: While MWI does not offer unique predictions that distinguish it from other interpretations, it provides a consistent and mathematically elegant framework for understanding quantum mechanics. Experimental tests primarily focus on verifying the predictions of quantum mechanics in general rather than specifically confirming MWI.

5. Criticisms and Controversies:

5.1 Ontological Complexity: Critics of MWI argue that the proliferation of parallel universes introduces unnecessary ontological complexity and violates Occam's razor, the principle of preferring simpler explanations. They contend that the interpretation lacks empirical evidence and remains speculative.

5.2 Subjective Experience: Some critics question the subjective experience of observers in MWI, arguing that it is difficult to reconcile the perception of a single definite outcome with the existence of multiple parallel branches. The subjective experience of observers within the multiverse remains a subject of philosophical debate.

Conclusion:

The Many-Worlds interpretation offers a provocative and unconventional perspective on the nature of reality, suggesting that our universe is just one of countless branches in a vast multiverse. While MWI has generated significant interest and debate among physicists and philosophers, it remains a speculative interpretation without direct empirical evidence. Whether MWI represents a fundamental truth about the nature of quantum reality or simply a mathematical formalism remains an open question, highlighting the ongoing exploration of the profound mysteries of quantum mechanics.

Pilot-Wave Theory and Other Interpretations

Pilot-wave theory, also known as de Broglie-Bohm theory, is an alternative interpretation of quantum mechanics that offers a deterministic view of reality. Developed by Louis de Broglie and David Bohm in the 1950s, pilot-wave theory posits the existence of a guiding wave that determines the trajectories of particles, while retaining the wave-particle duality of quantum mechanics. Here's an overview of pilot-wave theory and a comparison with other interpretations of quantum mechanics:

1. Pilot-Wave Theory:

1.1 Deterministic Dynamics: In pilot-wave theory, particles such as electrons are guided by a pilot wave, which evolves according to the Schrödinger equation. Unlike in the Copenhagen interpretation, where particles exist in a state of superposition until measured, particles in pilot-wave theory have well-defined trajectories at all times.

1.2 Hidden Variables: Pilot-wave theory introduces hidden variables, such as the positions of particles and the configuration of the guiding wave, which determine the behavior of quantum systems. These hidden variables account for the apparent randomness of quantum measurements and eliminate the need for wave function collapse.

1.3 Nonlocality: Pilot-wave theory preserves the nonlocal correlations predicted by quantum mechanics, allowing for instantaneous influences between distant particles. However, these nonlocal effects are deterministic and do not violate the principle of causality.

2. Comparison with Other Interpretations:

2.1 Copenhagen Interpretation: Pilot-wave theory contrasts with the Copenhagen interpretation in its rejection of wave function collapse and its deterministic description of particle trajectories. While the Copenhagen interpretation emphasizes the role of observers and the probabilistic nature of quantum mechanics, pilot-wave theory provides a deterministic realist alternative.

2.2 Many-Worlds Interpretation: In contrast to the Many-Worlds interpretation, which posits the existence of multiple parallel universes, pilot-wave theory maintains a single deterministic reality. It offers a more conservative interpretation of quantum mechanics, retaining a classical-like description of particles and waves.

2.3 Quantum Bayesianism (QBism): Quantum Bayesianism emphasizes the subjective nature of quantum probabilities and the role of observers in constructing their own personal probabilities. Pilot-wave theory, on the other hand, provides an objective, deterministic description of reality independent of observers' beliefs or perceptions.

3. Experimental Tests and Empirical Predictions:

3.1 Agreement with Quantum Mechanics: Pilot-wave theory reproduces the predictions of standard quantum mechanics for a wide range of phenomena, including the double-slit experiment, quantum tunneling, and quantum entanglement. It provides a mathematically consistent framework that is empirically equivalent to quantum mechanics.

3.2 Potential Tests: While pilot-wave theory has been primarily theoretical, recent experimental advances have enabled tests of its predictions. Experiments involving droplets on vibrating fluid surfaces, known as "pilot-wave experiments," have provided qualitative support for the underlying principles of pilot-wave theory.

Conclusion:

Pilot-wave theory offers a deterministic and realist interpretation of quantum mechanics, challenging the conventional wisdom of the Copenhagen interpretation and other probabilistic interpretations. While it remains a minority viewpoint within the physics community, pilot-wave theory provides an elegant and coherent framework for understanding the quantum world. Further experimental and theoretical research may shed light on the validity and implications of pilot-wave theory, contributing to our understanding of the fundamental nature of reality.

Chapter 13: Applications of Quantum Physics

Quantum physics has revolutionized our understanding of the fundamental laws governing the universe, and its practical applications have transformed technology, industry, and scientific research. In this chapter, we explore some of the most significant applications of quantum physics across various domains:

1. Quantum Computing:

1.1 Quantum Supremacy: Quantum computers harness the principles of quantum mechanics to perform complex calculations at speeds that surpass classical computers. Quantum supremacy, demonstrated by Google's Sycamore processor in 2019, showcases the potential of quantum computing to solve problems beyond the capabilities of classical systems.

1.2 Applications: Quantum computers have the potential to revolutionize fields such as cryptography, optimization, drug discovery, and machine learning. Quantum algorithms, such as Shor's algorithm for factoring large numbers and Grover's algorithm for database search, promise exponential speedup over classical counterparts.

2. Quantum Cryptography:

2.1 Unbreakable Encryption: Quantum cryptography utilizes the principles of quantum mechanics to achieve secure communication channels. Quantum key distribution (QKD) protocols enable the exchange of cryptographic keys with unconditional security, safeguarding against eavesdropping and information leakage.

2.2 Applications: Quantum cryptography has applications in secure communication networks, financial transactions, and government communications. Commercial implementations of QKD systems offer ultra-secure communication channels resistant to classical cryptographic attacks.

3. Quantum Sensing and Metrology:

3.1 Ultra-Precise Measurements: Quantum sensors leverage quantum phenomena such as superposition and entanglement to achieve unprecedented levels of precision in measurements. Examples include atomic clocks, magnetometers, and gravimeters that surpass the sensitivity of classical instruments.

3.2 Applications: Quantum sensing technologies have applications in navigation, geophysical exploration, medical imaging, and environmental monitoring. Quantum-enhanced sensors offer improved accuracy and sensitivity, enabling advancements in fields ranging from geology to healthcare.

4. Quantum Communication:

4.1 Secure Teleportation: Quantum communication protocols enable the teleportation of quantum states between distant locations, relying on the principles of entanglement and superposition. Quantum teleportation offers secure transmission of information and quantum states over long distances.

4.2 Applications: Quantum communication networks have applications in secure telecommunication, distributed computing, and quantum internet infrastructure. Research efforts aim to develop scalable quantum communication platforms for global quantum networks.

5. Quantum Materials and Technologies:

5.1 Superconductivity: Quantum materials exhibit exotic phenomena such as superconductivity, where electrical resistance vanishes at low temperatures. Superconducting materials find applications in MRI machines, particle accelerators, and quantum computing hardware.

5.2 Quantum Sensors and Detectors: Quantum technologies enable the development of ultra-sensitive detectors for detecting light, radiation, and other signals. Quantum sensors have applications in astronomy, medical imaging, and security screening.

Conclusion:

The applications of quantum physics span a wide range of fields, from information technology to healthcare and beyond. Quantum technologies have the potential to revolutionize industries, accelerate scientific discovery, and address complex challenges facing society. As research and development in quantum science continue to advance, the practical applications of quantum physics are poised to transform the way we live, work, and interact with the world around us.

Quantum Optics

Quantum optics is a branch of physics that explores the interaction between light and matter at the quantum level, incorporating principles from quantum mechanics and electromagnetism. It investigates phenomena such as the emission, absorption, and scattering of light by atoms and molecules, as well as the generation and manipulation of quantum states of light. Here's an overview of quantum optics and its key concepts:

1. Quantum Nature of Light:

1.1 Photon: In quantum optics, light is described as a collection of discrete particles called photons, each carrying a quantum of electromagnetic energy. Photons exhibit wave-particle duality, behaving as both particles and waves depending on the experimental context.
1.2 Wave Function: The quantum state of light is described by a wave function, which represents the probability amplitude for detecting photons at different positions and times. Quantum mechanics governs the evolution of the wave function and the behavior of photons in optical systems.

2. Atom-Photon Interactions:

2.1 Quantum Transitions: Atoms and molecules interact with light through processes such as absorption, spontaneous emission, and stimulated emission. These interactions involve quantum transitions between atomic energy levels, governed by selection rules and conservation laws.
2.2 Cavity Quantum Electrodynamics (QED): Cavity QED studies the interaction between atoms and photons confined within optical cavities. It explores phenomena such as vacuum fluctuations, photon trapping, and strong coupling between atoms and cavity modes.

3. Quantum States of Light:

3.1 Coherent States: Coherent states of light, such as laser light, exhibit well-defined phase and amplitude relationships and minimal quantum fluctuations. They are essential for applications such as interferometry, holography, and telecommunications.
3.2 Squeezed States: Squeezed states of light have reduced quantum fluctuations in one quadrature component at the expense of increased fluctuations in the orthogonal component. They find applications in quantum metrology, gravitational wave detection, and quantum information processing.

4. Quantum Optical Phenomena:

4.1 Quantum Entanglement: Quantum optics studies the generation, manipulation, and measurement of entangled states of light, where the properties of photons are correlated in nonclassical ways. Entanglement plays a crucial role in quantum communication, teleportation, and cryptography.

4.2 Single-Photon Sources: Quantum optics research focuses on developing sources of single photons with high purity, indistinguishability, and coherence. Single-photon sources are fundamental for applications in quantum computing, quantum cryptography, and quantum imaging.

5. Quantum Optical Technologies:

5.1 Quantum Information Processing: Quantum optics is instrumental in the development of quantum information processing technologies, including quantum gates, quantum memories, and quantum repeaters. These technologies enable the manipulation and transmission of quantum information for computation and communication tasks.

5.2 Quantum Metrology: Quantum optics techniques provide ultra-sensitive tools for precision measurement and metrology, surpassing the limits of classical optical instruments. Quantum metrology has applications in timekeeping, navigation, and fundamental physics research.

Conclusion:

Quantum optics lies at the intersection of quantum mechanics, electromagnetism, and photonics, offering insights into the fundamental nature of light and its interaction with matter. From the study of atom-photon interactions to the development of quantum optical technologies, quantum optics plays a crucial role in advancing scientific understanding and enabling practical applications in diverse fields. As research in quantum optics continues to progress, it promises to unlock new capabilities for information processing, sensing, and communication, shaping the future of technology and science.

Quantum Sensing and Metrology

Quantum sensing and metrology are fields that leverage the principles of quantum mechanics to achieve ultra-sensitive measurements of physical quantities such as time, distance, magnetic fields, and gravitational fields. These technologies enable advancements in precision measurement, navigation, imaging, and fundamental research. Let's explore quantum sensing and metrology in more detail:

1. Quantum Sensing:

1.1 Principles: Quantum sensors utilize quantum phenomena such as superposition and entanglement to achieve unprecedented levels of precision in measuring physical quantities. By exploiting quantum effects, these sensors can surpass the sensitivity of classical sensors and detect subtle signals with high accuracy.

1.2 Examples:

- Atomic Clocks: Atomic clocks measure time by counting the oscillations of atoms, such as cesium or rubidium atoms. Quantum effects, such as the hyperfine transition in cesium, provide stable and precise frequency references for timekeeping.
- Magnetometers: Quantum magnetometers measure magnetic fields with high sensitivity by exploiting the quantum properties of atoms or molecules. Examples include optically pumped magnetometers and atomic magnetometers based on spin-exchange relaxation.

2. Quantum Metrology:

2.1 Fundamentals: Quantum metrology aims to enhance the precision of measurements beyond the limits imposed by classical physics. By harnessing quantum resources such as entanglement and squeezed states, quantum metrology enables measurements with reduced uncertainty and improved sensitivity.

2.2 Examples:
- Gravitational Wave Detectors: Gravitational wave detectors such as LIGO (Laser Interferometer Gravitational-Wave Observatory) and VIRGO use interferometric techniques to detect minuscule distortions in spacetime caused by passing gravitational waves. Quantum-enhanced techniques improve the sensitivity of these detectors, enabling the detection of fainter signals.
- Quantum Imaging: Quantum metrology techniques can enhance the resolution and sensitivity of imaging systems, allowing for the detection of low-light or low-contrast objects. Quantum imaging has applications in medical imaging, remote sensing, and security screening.

3. Applications:

3.1 Navigation and Positioning: Quantum sensors offer improved accuracy and reliability for navigation systems such as GPS (Global Positioning System). Quantum gyroscopes and accelerometers provide precise measurements of rotation and acceleration, essential for inertial navigation systems in aircraft, ships, and satellites.

3.2 Geophysical Exploration: Quantum sensors enable the detection and mapping of subsurface structures, mineral deposits, and geological formations. Quantum magnetometers and gravimeters have applications in mineral exploration, oil and gas prospecting, and environmental monitoring.

3.3 Medical Diagnostics: Quantum sensing technologies can improve the sensitivity and specificity of medical diagnostic tools, leading to earlier detection of diseases and more effective treatments. Quantum-enhanced imaging techniques enable high-resolution imaging of biological tissues and organs.

Conclusion:

Quantum sensing and metrology represent cutting-edge technologies with applications across various domains, from fundamental research to practical applications in industry and healthcare. By harnessing the unique properties of quantum mechanics, these technologies enable precise measurements of physical quantities with unprecedented sensitivity and accuracy. As research in quantum sensing and metrology continues to advance, these technologies hold the promise of revolutionizing fields such as navigation, geophysics, medical diagnostics, and beyond, opening up new frontiers in scientific exploration and technological innovation.

Quantum Biology

Quantum biology is an interdisciplinary field that explores the role of quantum mechanics in biological processes, ranging from the molecular scale to entire organisms. While classical physics adequately describes many biological phenomena, quantum mechanics becomes essential for understanding processes that involve light-harvesting, energy transfer, and molecular interactions at the quantum level. Here's an overview of quantum biology and its key concepts:

1. Photosynthesis:

1.1 Light Harvesting: Photosynthetic organisms, such as plants, algae, and certain bacteria, utilize specialized pigment-protein complexes to capture and convert light energy into chemical energy. Quantum coherence and quantum superposition play crucial roles in optimizing the efficiency of light-harvesting and energy transfer processes.
1.2 Exciton Transport: Excitons, which are quasi-particles representing electronic excitations, can exhibit quantum coherence and delocalization over multiple chromophores in photosynthetic complexes. Quantum coherence enhances the efficiency of energy transfer by enabling excitons to explore multiple pathways simultaneously.

2. Enzyme Catalysis:

2.1 Hydrogen Tunneling: Enzymes catalyze biochemical reactions by lowering the activation energy barrier for chemical transformations. Quantum tunneling, a quantum mechanical phenomenon, allows protons and electrons to "tunnel" through energy barriers, facilitating enzymatic reactions such as proton transfer and hydrogen exchange.
2.2 Proton-Coupled Electron Transfer: Quantum mechanics governs proton-coupled electron transfer reactions, where the transfer of both protons and electrons occurs concurrently. These reactions are fundamental to processes such as respiration, photosynthesis, and DNA repair.

3. Magnetoreception:

3.1 Magnetic Navigation: Certain organisms, including birds, insects, and marine animals, navigate using the Earth's magnetic field—a phenomenon known as magnetoreception. Quantum effects, such as the radical pair mechanism, may underlie the detection and processing of magnetic signals in specialized photoreceptor proteins.

3.2 Cryptochrome Proteins: Cryptochromes are blue-light-sensitive photoreceptor proteins found in plants and animals, including birds and insects. Quantum coherence and the radical pair mechanism in cryptochromes have been proposed to play a role in magnetoreception and the sensing of magnetic fields.

4. Quantum Entanglement:

4.1 Entangled States: Quantum entanglement, where the quantum states of particles become correlated in such a way that the state of one particle is dependent on the state of another, may have implications for biological systems. Some researchers speculate that entanglement could be involved in long-range communication or information processing within living organisms.
4.2 Biological Signaling: While direct evidence of entanglement in biological systems remains elusive, theoretical studies suggest that entangled states could enhance the efficiency of signaling and communication in neuronal networks or cellular pathways.

5. Quantum Sensors in Biology:

5.1 Quantum Probes: Quantum sensors and imaging techniques offer new opportunities for probing biological structures and processes with high sensitivity and resolution. Quantum-enhanced sensors, such as atomic magnetometers and quantum dots, enable non-invasive imaging and detection of biological samples with minimal perturbation.
5.2 Biological Imaging: Quantum-based imaging modalities, including quantum dots, fluorescence resonance energy transfer (FRET), and quantum-enhanced microscopy, provide valuable tools for visualizing cellular structures, molecular interactions, and dynamic processes in living organisms.

Conclusion:

Quantum biology bridges the gap between quantum physics and biology, offering insights into the fundamental principles underlying biological systems and processes. By integrating concepts from quantum mechanics with experimental and theoretical approaches in biology, quantum biology provides a deeper understanding of how living organisms harness quantum phenomena to perform essential functions such as energy conversion, biochemical reactions, and sensory perception. Continued research in quantum biology promises to unravel new mysteries of life and inspire novel approaches to biotechnology, medicine, and environmental science.

Chapter 14: Future Directions in Quantum Physics

Quantum physics has led to groundbreaking discoveries and technological advancements, but many exciting avenues of research and exploration lie ahead. In this chapter, we discuss the future directions and emerging trends in quantum physics, including theoretical developments, experimental breakthroughs, and applications in various fields:

1. Quantum Computing:

1.1 Fault-Tolerant Quantum Computers: Future research aims to overcome the challenges of decoherence and error correction to build fault-tolerant quantum computers capable of solving complex problems beyond the reach of classical computers.
1.2 Quantum Algorithms: Continued development of quantum algorithms, such as quantum machine learning, optimization, and cryptography, promises to unlock the full potential of quantum computing for practical applications in science, engineering, and industry.

2. Quantum Communication and Networking:

2.1 Quantum Internet: Efforts are underway to develop a quantum internet infrastructure that enables secure communication, distributed quantum computing, and quantum-enhanced sensing on a global scale.
2.2 Quantum Key Distribution (QKD): Advances in quantum cryptography will lead to the widespread deployment of QKD systems for ultra-secure communication, protecting sensitive data against eavesdropping and cyberattacks.

3. Quantum Sensing and Metrology:

3.1 Quantum Sensors: Future quantum sensors will push the limits of sensitivity and resolution, enabling precise measurements of gravitational waves, magnetic fields, biological signals, and environmental parameters.
3.2 Quantum Metrology: Quantum-enhanced metrology techniques will revolutionize fields such as timekeeping, navigation, geodesy, and fundamental constants, leading to improved accuracy and reliability in measurement standards.

4. Quantum Materials and Technologies:

4.1 Quantum Materials: Advances in the synthesis and characterization of quantum materials will lead to the discovery of novel properties and applications, including topological insulators, superconductors, and quantum spin liquids.
4.2 Quantum Devices: Quantum-enhanced devices, such as quantum sensors, quantum memories, and quantum processors, will become increasingly integrated into technological platforms, driving innovation in electronics, photonics, and information technology.

5. Quantum Biology and Quantum Neuroscience:

5.1 Biological Applications: Quantum biology research will continue to uncover the role of quantum phenomena in biological systems, from photosynthesis and enzyme catalysis to sensory perception and cognition.
5.2 Brain-Computer Interfaces: Quantum-inspired approaches to neuroscience and brain-computer interfaces may lead to new insights into brain function, consciousness, and the development of advanced neurotechnologies for healthcare and human augmentation.

Conclusion:

The future of quantum physics holds immense promise for scientific discovery, technological innovation, and societal impact. As researchers push the boundaries of knowledge and technology, quantum physics will continue to revolutionize our understanding of the universe and empower us to address complex challenges in areas such as computation, communication, sensing, and healthcare. By fostering collaboration between scientists, engineers, and policymakers, we can harness the full potential of quantum physics to shape a brighter future for humanity.

Quantum Technologies on the Horizon

Quantum technologies represent a paradigm shift in science and engineering, offering unprecedented capabilities for computation, communication, sensing, and imaging. As these technologies continue to advance, several promising developments are on the horizon:

1. Fault-Tolerant Quantum Computing:

1.1 Error Correction: Research efforts are focused on developing error-correcting codes and fault-tolerant architectures to mitigate the effects of decoherence and noise in quantum computers, enabling reliable operation at scale.
1.2 Scalable Platforms: Novel qubit technologies, including trapped ions, superconducting circuits, and topological qubits, are being pursued to create scalable quantum computing platforms capable of solving complex problems with practical applications.

2. Quantum Communication Networks:

2.1 Quantum Repeaters: Development of quantum repeater protocols will extend the range and efficiency of quantum communication networks, enabling secure transmission of quantum information over long distances and across multiple nodes.
2.2 Quantum Satellite Communication: Implementation of quantum communication protocols via satellite-based platforms will provide secure global communication links immune to interception or tampering, advancing secure telecommunications and data exchange.

3. Quantum Sensing and Imaging:

3.1 High-Precision Sensors: Quantum-enhanced sensors, such as atomic magnetometers and quantum gravimeters, will offer unparalleled sensitivity and accuracy for detecting magnetic fields, gravitational waves, and other physical quantities, with applications in geophysics, healthcare, and defense.

3.2 Quantum Imaging Technologies: Quantum-inspired imaging techniques, including quantum-enhanced microscopy and quantum radar, will enable high-resolution imaging and sensing capabilities for biomedical imaging, remote sensing, and non-destructive testing.

4. Quantum Materials and Devices:

4.1 Quantum Materials Engineering: Discovery and manipulation of novel quantum materials, such as topological insulators, Majorana fermions, and quantum spin liquids, will lead to the development of advanced electronic, photonic, and spintronic devices with unique properties and functionalities.

4.2 Quantum Devices Integration: Integration of quantum devices, such as quantum dots, single-photon sources, and quantum memories, into existing technologies will pave the way for quantum-enhanced devices and systems with enhanced performance and functionality.

5. Quantum Biology and Healthcare:

5.1 Biological Sensing and Imaging: Quantum-inspired techniques for biological sensing and imaging will enable non-invasive, high-resolution imaging of biological structures and processes, advancing diagnostics, drug discovery, and personalized medicine.

5.2 Neuromorphic Computing: Development of neuromorphic computing architectures inspired by quantum principles will facilitate the simulation of complex biological systems and neural networks, leading to breakthroughs in artificial intelligence and brain-computer interfaces.

Conclusion:

The horizon of quantum technologies holds immense promise for transforming industries, revolutionizing scientific research, and addressing societal challenges. As researchers and engineers continue to push the boundaries of quantum science and engineering, we can anticipate a future where quantum technologies play an increasingly integral role in shaping our world, driving innovation, and fostering a new era of discovery and progress.

Challenges and Opportunities

Quantum technologies offer unprecedented opportunities for revolutionizing computation, communication, sensing, and beyond. However, they also present several challenges that must be addressed to realize their full potential. Let's explore some of these challenges and opportunities:

Challenges:

1. Decoherence and Error Correction: Quantum systems are highly susceptible to decoherence and errors caused by environmental noise and imperfections in hardware. Developing robust error correction techniques and fault-tolerant architectures is essential for building reliable quantum computers and communication networks.
2. Scalability: Scaling up quantum systems to accommodate large numbers of qubits and complex operations remains a significant challenge. Overcoming scalability limitations requires advances in qubit coherence times, control methods, and physical integration of quantum components.
3. Resource Requirements: Quantum technologies often require specialized infrastructure, including ultra-low temperatures, high vacuum environments, and precise control instrumentation. Addressing the resource requirements and operational constraints of quantum systems is essential for practical implementation and widespread adoption.
4. Interfacing with Classical Systems: Integrating quantum technologies with existing classical systems poses challenges in terms of compatibility, interface protocols, and data transfer. Bridging the gap between quantum and classical computing, communication, and sensing is critical for seamless integration and interoperability.

Opportunities:

1. Unprecedented Computing Power: Quantum computers promise exponential speedup for solving complex computational problems, including optimization, cryptography, and material simulation. Harnessing this computing power will enable breakthroughs in scientific research, engineering design, and data analysis.
2. Secure Communication: Quantum communication offers unbreakable encryption and secure transmission of sensitive information, safeguarding against cyber threats and privacy breaches. Quantum key distribution and secure quantum networks provide opportunities for enhancing cybersecurity and protecting digital infrastructure.
3. Ultra-Sensitive Sensing and Imaging: Quantum sensors and imaging techniques offer unparalleled sensitivity and resolution for detecting physical quantities, biological signals, and imaging structures at the nanoscale. Quantum-enhanced sensing has applications in healthcare, environmental monitoring, and defense.
4. Advanced Materials and Devices: Discovering and engineering novel quantum materials and devices enable the development of next-generation electronics, photonics, and energy technologies. Quantum-inspired devices offer unique functionalities, such as spintronics, quantum dots, and quantum memory, with applications in information processing, sensing, and quantum computing.

Conclusion:

While quantum technologies present formidable challenges, they also offer transformative opportunities for innovation, discovery, and societal impact. Overcoming the challenges of decoherence, scalability, and resource requirements will require concerted efforts from researchers, engineers, policymakers, and industry stakeholders. By addressing these challenges and seizing the opportunities presented by quantum technologies, we can unlock new frontiers in science, technology, and human endeavor, shaping a future where quantum capabilities empower us to address complex challenges and create a more secure, sustainable, and interconnected world.

The Future of Quantum Research

The future of quantum research is rich with possibilities, driven by ongoing advancements in theory, experimentation, and technological innovation. Here are some key areas that are likely to shape the future trajectory of quantum research:

1. Quantum Computing:

1.1 Beyond NISQ: As quantum computers evolve beyond the noisy intermediate-scale quantum (NISQ) era, researchers will focus on building fault-tolerant, error-corrected quantum computers capable of solving real-world problems with practical applications.
1.2 Quantum Algorithms: Development of new quantum algorithms and quantum programming languages will enable the efficient implementation of quantum solutions for optimization, cryptography, machine learning, and material science.

2. Quantum Communication and Networking:

2.1 Global Quantum Internet: Research efforts will continue to advance the development of a secure, scalable quantum internet infrastructure, enabling long-distance quantum communication, distributed quantum computing, and quantum-enhanced sensing.
2.2 Quantum Repeaters: Improvements in quantum repeater technology will extend the range and efficiency of quantum communication networks, facilitating secure quantum communication over intercontinental distances.

3. Quantum Sensing and Metrology:

3.1 Ultra-Precise Sensors: Quantum sensing technologies will achieve unprecedented levels of sensitivity and resolution, enabling high-precision measurements of physical quantities such as magnetic fields, gravitational waves, and biological signals.

3.2 Quantum Metrology: Advances in quantum-enhanced metrology techniques will lead to the development of next-generation measurement standards, enabling improved accuracy and reliability in timekeeping, navigation, and fundamental constants.

4. Quantum Materials and Devices:

4.1 Discovery and Engineering: Research into quantum materials will uncover new phenomena and properties, leading to the discovery of novel materials with unique quantum functionalities for electronics, photonics, and energy applications.
4.2 Quantum Devices Integration: Integration of quantum devices into existing technology platforms will enable the development of quantum-enhanced devices and systems with improved performance and functionality, driving innovation in information processing, sensing, and computing.

5. Quantum Biology and Quantum Neuroscience:

5.1 Understanding Complexity: Quantum biology research will deepen our understanding of the role of quantum phenomena in biological systems, from photosynthesis and enzyme catalysis to sensory perception and cognition, providing insights into the fundamental mechanisms of life.
5.2 Neuromorphic Computing: Exploration of quantum-inspired approaches to neuroscience and brain-inspired computing will lead to the development of advanced neuromorphic computing architectures and brain-computer interfaces, with applications in artificial intelligence and healthcare.

Conclusion:

The future of quantum research is characterized by a convergence of interdisciplinary efforts spanning physics, engineering, biology, and information science. By addressing fundamental scientific challenges and translating theoretical insights into practical applications, quantum research holds the promise of unlocking new frontiers in technology, enabling transformative innovations, and addressing pressing societal challenges. With sustained investment, collaboration, and exploration, the future of quantum research is boundless, offering endless possibilities for discovery, exploration, and innovation.

Chapter 15: Conclusion: Embracing the Quantum World

The journey through the realm of quantum physics has been one of awe-inspiring discovery, revealing the astonishing nature of reality at its most fundamental level. As we conclude our exploration, it is clear that embracing the quantum world opens doors to endless possibilities and transformative potential. Here are some key reflections on the significance of quantum physics and its implications for the future:

1. Understanding the Quantum Universe:

1.1 Beyond Intuition: Quantum physics challenges our classical intuition, revealing a world where particles can exist in multiple states simultaneously, where measurements influence outcomes, and where randomness and uncertainty reign supreme.
1.2 Unified Framework: Quantum mechanics provides a unified framework for understanding the behavior of particles, waves, and fields, encompassing phenomena from the microscopic realm of atoms and subatomic particles to the cosmic scales of the universe.

2. Quantum Technologies and Innovations:

2.1 Transformative Potential: Quantum technologies hold the promise of revolutionizing computing, communication, sensing, and imaging, offering unparalleled capabilities for solving complex problems, securing information, and advancing scientific discovery.
2.2 Real-World Applications: From quantum computers and cryptography to quantum sensors and materials, the practical applications of quantum physics are poised to reshape industries, drive innovation, and address societal challenges in healthcare, energy, and beyond.

3. Embracing Quantum Principles:

3.1 Harnessing Quantum Phenomena: Embracing the principles of quantum mechanics enables us to harness the power of superposition, entanglement, and coherence to develop new technologies, explore the frontiers of science, and unlock deeper insights into the nature of reality.
3.2 Interdisciplinary Collaboration: Quantum physics transcends disciplinary boundaries, inviting collaboration between physicists, engineers, biologists, and computer scientists to tackle complex problems and explore the intersection of quantum theory with other fields of inquiry.

4. Ethical and Societal Considerations:

4.1 Ethical Implications: As quantum technologies become more prevalent, it is essential to consider their ethical implications, including issues of privacy, security, and equity, and to ensure responsible development and deployment.
4.2 Global Collaboration: Embracing the quantum world requires global collaboration and cooperation, fostering open exchange of knowledge, resources, and expertise to address common challenges and realize the full potential of quantum technologies for the benefit of humanity.

5. Embracing the Unknown:

5.1 Exploration and Discovery: The journey through the quantum world is one of exploration and discovery, where each new revelation opens up new avenues of inquiry and invites us to confront the mysteries of the universe with humility and curiosity.

5.2 Infinite Possibilities: As we embrace the quantum world, we embark on a journey of endless possibilities, where the boundaries of what is known and what is possible continue to expand, inspiring wonder, imagination, and the pursuit of knowledge.

In conclusion, embracing the quantum world is a profound invitation to explore the frontiers of science, technology, and human understanding. As we navigate the complexities of the quantum realm, let us approach this journey with a spirit of curiosity, creativity, and collaboration, embracing the mysteries of the universe and the transformative potential of quantum physics to shape a better future for generations to come.

Recapitulation of Key Concepts

Throughout our exploration of quantum physics, we have encountered several key concepts that underpin our understanding of the quantum world. Let's recapitulate these fundamental ideas:

1. Wave-Particle Duality:

1.1 Dual Nature: Particles such as electrons and photons exhibit both wave-like and particle-like behavior, as demonstrated by phenomena like diffraction, interference, and photoelectric effect.
1.2 Quantum Superposition: Quantum objects can exist in multiple states simultaneously, represented by a superposition of wavefunctions, until measured.

2. Uncertainty Principle:

2.1 Heisenberg's Principle: The uncertainty principle states that certain pairs of physical properties, such as position and momentum, cannot be precisely determined simultaneously beyond a certain limit.
2.2 Implications: Uncertainty places fundamental limits on our ability to predict and measure the behavior of quantum systems, giving rise to probabilistic interpretations of quantum mechanics.

3. Quantum States and Entanglement:

3.1 State Vectors: Quantum states are described by complex-valued wavefunctions or state vectors, representing the probabilities of finding a particle in various states upon measurement.
3.2 Entanglement: Entanglement occurs when the quantum states of two or more particles become correlated in such a way that the state of one particle is dependent on the state of the others, even when separated by large distances.

4. Quantum Mechanics:

4.1 Schrödinger Equation: The Schrödinger equation governs the evolution of quantum states over time, describing how wavefunctions propagate and evolve in response to external potentials.

4.2 Operators and Observables: Operators represent physical observables in quantum mechanics, and their eigenvalues correspond to the possible outcomes of measurements.

5. Quantum Computing and Cryptography:

5.1 Quantum Computing: Quantum computers leverage quantum superposition and entanglement to perform parallel computations, potentially offering exponential speedup over classical computers for certain tasks.

5.2 Quantum Cryptography: Quantum cryptography enables secure communication using quantum key distribution protocols, leveraging the principles of quantum mechanics to ensure the confidentiality and integrity of transmitted data.

6. Quantum Fields and Particles:

6.1 Quantization of Fields: Quantum field theory describes fields as quantized excitations, giving rise to particles as quanta of these fields, such as photons for the electromagnetic field and electrons for the electron field.

6.2 Interactions and Forces: Fundamental forces in nature, such as electromagnetism, weak force, and strong force, are mediated by exchange particles, which are manifestations of underlying quantum fields.

Conclusion:

The concepts of quantum physics encompass a rich tapestry of ideas that challenge our classical intuition and provide a deeper understanding of the fundamental laws governing the universe. By grappling with these key concepts, we gain insights into the mysterious and wondrous nature of quantum reality, opening up new horizons for scientific exploration, technological innovation, and philosophical inquiry.

Importance of Quantum Physics in Modern Science and Technology

The importance of quantum physics in modern science and technology cannot be overstated. Quantum mechanics, with its revolutionary concepts and principles, has transformed our understanding of the natural world and enabled the development of groundbreaking technologies. Here are some key reasons why quantum physics is indispensable in modern science and technology:

1. Quantum Computing:

1.1 Exponential Speedup: Quantum computers have the potential to solve certain computational problems exponentially faster than classical computers, offering breakthroughs in areas such as cryptography, optimization, drug discovery, and materials science.
1.2 Quantum Algorithms: Quantum algorithms, such as Shor's algorithm for factoring large numbers and Grover's algorithm for database search, leverage quantum principles to perform tasks more efficiently than classical counterparts.

2. Quantum Communication and Cryptography:

2.1 Unbreakable Encryption: Quantum cryptography provides unbreakable encryption methods based on the principles of quantum mechanics, ensuring secure communication channels resistant to eavesdropping and hacking.
2.2 Quantum Key Distribution: Quantum key distribution (QKD) enables the exchange of cryptographic keys with unconditional security, offering a quantum-safe solution for protecting sensitive data in the era of quantum computing.

3. Quantum Sensing and Metrology:

3.1 Ultra-Precise Measurements: Quantum sensors and metrology techniques offer unparalleled sensitivity and accuracy for detecting physical quantities such as magnetic fields, gravitational waves, and biological signals, with applications in healthcare, environmental monitoring, and fundamental physics research.
3.2 High-Resolution Imaging: Quantum-enhanced imaging technologies enable high-resolution imaging of biological tissues, materials, and structures at the nanoscale, advancing fields such as medical diagnostics, materials science, and non-destructive testing.

4. Quantum Materials and Devices:

4.1 Novel Functionalities: Quantum materials exhibit unique properties and functionalities, such as superconductivity, topological insulators, and quantum spin liquids, leading to the development of advanced electronic, photonic, and energy devices with transformative applications.
4.2 Quantum Devices Integration: Integration of quantum devices, such as quantum dots, single-photon sources, and quantum memories, into existing technology platforms enables the creation of quantum-enhanced devices and systems for information processing, communication, and computing.

5. Fundamental Science and Understanding:

5.1 Fundamental Laws of Nature: Quantum physics provides the theoretical framework for understanding the behavior of particles, fields, and interactions at the smallest scales, shedding light on the fundamental laws that govern the universe.

5.2 New Frontiers of Exploration: Quantum research opens up new frontiers of scientific exploration, challenging our classical worldview and inspiring discoveries in areas such as quantum gravity, quantum biology, and quantum cosmology.

Conclusion:

In summary, quantum physics lies at the heart of modern science and technology, driving innovations that have the potential to revolutionize society and reshape the future. By harnessing the principles of quantum mechanics, researchers and engineers are pushing the boundaries of what is possible, unlocking new capabilities and insights that promise to transform our world in profound and unexpected ways. As we continue to explore the quantum realm, the importance of quantum physics in modern science and technology will only grow, paving the way for a quantum-enabled future filled with endless possibilities.

The Beauty and Mystery of the Quantum Realm

The beauty and mystery of the quantum realm captivate scientists, philosophers, and dreamers alike, inviting us to explore the enigmatic and wondrous nature of reality at its most fundamental level. Here, in the ethereal realm of quantum physics, we encounter a tapestry of concepts and phenomena that challenge our understanding and ignite our imagination. Let us delve into the captivating allure of the quantum realm:

1. Wave-Particle Duality:

In the quantum realm, particles exhibit a dual nature, behaving both as discrete particles and as waves with wave-like properties. This duality, as exemplified by the famous double-slit experiment, showcases the inherent ambiguity and richness of quantum phenomena, inviting us to ponder the nature of existence itself.

2. Uncertainty and Probability:

Uncertainty reigns supreme in the quantum realm, where the precise properties of particles, such as their position and momentum, elude deterministic prediction. Instead, we are confronted with a realm of probabilities, where the outcome of measurements is governed by probabilistic distributions, challenging our classical notions of causality and determinism.

3. Quantum Entanglement:

Entanglement, perhaps the most mystifying aspect of quantum physics, reveals the interconnectedness and non-locality of quantum systems. When particles become entangled, their states become intimately linked, regardless of the distance between them, defying our classical intuition and suggesting hidden layers of reality waiting to be unveiled.

4. Superposition and Coherence:

Superposition, the hallmark of quantum mechanics, allows particles to exist in multiple states simultaneously, embodying a state of potentiality and ambiguity. Coherence, the delicate maintenance of quantum states, reveals the fragility and beauty of quantum systems, where interference patterns emerge and vanish in a dance of waves and particles.

5. Quantum Weirdness:

The quantum realm is replete with paradoxes and counterintuitive phenomena that challenge our comprehension and stretch the boundaries of our imagination. From particles tunneling through barriers to observers influencing the behavior of quantum systems, the quantum world delights and confounds with its inherent strangeness and unpredictability.

Embracing the Mystery:

In the face of such beauty and mystery, we are compelled to embrace the quantum realm with humility and wonder, recognizing the limitations of our classical intuitions and the boundless possibilities that lie beyond. As we journey deeper into the quantum landscape, let us revel in the elegance of its mathematical formalism, the poetry of its conceptual frameworks, and the profound insights it offers into the nature of reality.

Conclusion:

The beauty and mystery of the quantum realm beckon us to explore, to question, and to marvel at the wonders of the universe. In this realm of uncertainty and possibility, we encounter not only the frontiers of science but also the depths of human imagination and curiosity. As we navigate the intricacies of the quantum world, let us embrace its mysteries with reverence and awe, for it is in the pursuit of understanding that we uncover the true essence of beauty and wonder in the cosmos.

Appendix: Glossary of Quantum Terms

1. Wave-Particle Duality: The concept that particles, such as electrons and photons, exhibit both wave-like and particle-like behavior.
2. Uncertainty Principle: A fundamental principle of quantum mechanics, formulated by Werner Heisenberg, stating that certain pairs of physical properties, such as position and momentum, cannot be precisely determined simultaneously.
3. Quantum Superposition: The principle that quantum objects can exist in multiple states simultaneously until measured, represented by a linear combination of wavefunctions.
4. Quantum Entanglement: A phenomenon where the quantum states of two or more particles become correlated in such a way that the state of one particle is dependent on the state of the others, even when separated by large distances.
5. Schrödinger Equation: The fundamental equation of quantum mechanics, describing how the wavefunction of a quantum system evolves over time in response to external potentials.
6. Quantum Computing: A computing paradigm that harnesses quantum phenomena such as superposition and entanglement to perform calculations more efficiently than classical computers for certain tasks.
7. Quantum Cryptography: Cryptographic techniques based on the principles of quantum mechanics, offering unbreakable encryption methods for secure communication.
8. Quantum Sensing: Sensing techniques that leverage quantum principles to achieve high sensitivity and accuracy in detecting physical quantities such as magnetic fields, gravitational waves, and biological signals.
9. Quantum Materials: Materials with unique quantum properties and functionalities, such as superconductivity, topological insulators, and quantum spin liquids.
10. Quantum Field Theory: A theoretical framework that combines quantum mechanics with classical field theory, describing the behavior of quantum fields and particles as excitations of these fields.
11. Quantum Electrodynamics (QED): A quantum field theory that describes the interaction of electrons and photons, providing a framework for understanding electromagnetic phenomena at the quantum level.
12. Quantum Chromodynamics (QCD): A quantum field theory that describes the strong force, which binds quarks together to form protons, neutrons, and other hadrons.
13. Quantum Gravity: Theoretical attempts to describe gravity within the framework of quantum mechanics, aiming to unify the fundamental forces of nature.
14. Quantum Cosmology: The application of quantum principles to the study of the origin, evolution, and fate of the universe, addressing questions about the nature of space, time, and cosmological constants.
15. Quantum Interpretations: Different interpretations of quantum mechanics, such as the Copenhagen interpretation, Many-Worlds interpretation, and Pilot-Wave theory, which offer different perspectives on the meaning of quantum phenomena.

This glossary provides a brief overview of key terms and concepts in quantum physics, serving as a reference for readers seeking to deepen their understanding of the quantum realm.